Contributions to Management Science

The series *Contributions to Management Science* contains research publications in all fields of business and management science. These publications are primarily monographs and multiple author works containing new research results, and also feature selected conference-based publications are also considered. The focus of the series lies in presenting the development of latest theoretical and empirical research across different viewpoints.

This book series is indexed in Scopus.

Chen Zhang • Yu Gong • Steve Brown

Blockchain Applications in Food Supply Chain Management

Case Studies and Implications

Springer

Chen Zhang
University of Southampton
Southampton, UK

Yu Gong
University of Southampton
Southampton, UK

Steve Brown
University of Sussex
Brighton, UK

ISSN 1431-1941 ISSN 2197-716X (electronic)
Contributions to Management Science
ISBN 978-3-031-27053-6 ISBN 978-3-031-27054-3 (eBook)
https://doi.org/10.1007/978-3-031-27054-3

This Springer imprint is published by the registered company Springer Nature Switzerland AG
The registered company address is: Gewerbestrasse 11, 6330 Cham, Switzerland

Contents

List of Figures

List of Tables

Chapter 1
Introduction

In this chapter, the project background is introduced. This includes the motivation for this research and how the paper can contribute to the broader field of study. The current food system is complex, and food hazards are happening more frequently than ever. Therefore, the first section introduces food quality and safety issues in the current food industry. This section presents the food industry reality with real-world examples. The second section describes the food supply chain and its importance to the food industry. The third part discusses the issue of traceability and its contribution to food quality and safety. The following section presents some current traceability innovations. Blockchain is introduced in the fifth section with lists of real-world application examples. The last section proposes the research questions and sets out the structure of the research.

1.1 Food System

Food systems are complex and keep changing over time (Yakovleva, 2007; Wognum et al., 2011). There are many issues in the current food system, including climate change, foodborne diseases, food shortage, malnutrition, resources insufficiency, and food wastage, among others. Among all the issues, food quality and safety issues are gaining more attention due to an increasing trend of food hazards in recent years. Food safety is a major concern when customers are making purchasing decisions (Sims, 2018). Before the 1980s, food quantity was the major issue (Kaferstein, 2003). The food safety issue was first properly addressed in 1983 by the first Expert Committee that was held by World Trade Organization (WTO) and the Food and Agriculture Organization (FAO). Accessing nutritionally adequate and safe food is a basic human right, according to the joint FAO/WHO on International Conference of Nutrition in 1992. However, to satisfy the individual's right of food safety is more of a promising idea than a reality (Kaferstein, 2003).

C. Zhang et al., *Blockchain Applications in Food Supply Chain Management*, Contributions to Management Science,
https://doi.org/10.1007/978-3-031-27054-3_1

In terms of food hazards, unsafe foods can cause more than 200 diseases, with almost 1 in 10 people becoming sick each year (WHO, 2020). More specifically, as one of the counties with the safest food supply, the US still has 48 million cases of foodborne illness every year, which can result in about 3000 deaths and 128,000 hospitalisations, according to the FDA (2019a). In 2008 and 2009, nine people died and at least 714 people were infected from eating products containing salmonella-contaminated peanuts. It was one of the largest food disease issues in the USA's history (Steenhuysen, 2009). The inadequate level of cleanliness in the producing factory included mouldy ceilings and cockroaches, and the lack of a handwashing sink was reported (Steenhuysen, 2009) as a cause. In other developed areas, the horse meat scandal was revealed in the UK in 2013. The investigation found undeclared horse meat in beef burgers (Telegraph, 2013). Most recently, E. coli-contaminated lettuce has caused at least five deaths in the USA and spread across 35 states in the USA (Wheaton, 2018).

Compared to developed countries, the number of deaths and illnesses caused by food hazards in developing regions can only be higher due to multiple factors such as: population growth, lack of health-related infrastructures, lack of policies, poverty, and lack of hygiene education (Kaferstein, 2003; Aruoma, 2006). One of the most shocking food incidents in Chinese food history happened in 2008. One of the biggest national diary companies—Sanlu—was accused of using the toxic chemical melamine in milk powder to increase the protein content, this caused at least six babies' deaths and harmed the health of thousands of children (BBC, 2010). To increase the protein content, milk powder producers used melamine, which is a toxic ingredient normally found in plastic, fertilisers and cleaning products, to replace protein. After the incident was revealed, 19 people who were working in the factories were jailed for using toxic ingredients. The executives of Sanlu companies were also arrested because of selling toxic products (BBC, 2010; Business & Human rights resource centre, 2010). According to a survey in 2011, food safety was the top issue that concerned Chinese people (Lam et al., 2013). There were 6685 food incidents reported officially in China in 2012 (Lam et al., 2013).

The food trade is now one of the largest global businesses, and the food system is becoming even more complex with more stakeholders involved (Roth et al., 2008; Christopher et al., 2011; Storoy et al., 2013; Ringsberg, 2014; Mattevi & Jones, 2016). It is very common to have one food product with a combination of multiple ingredients sourced from various countries, as Roth et al. (2008) proposed that global sourcing of food and ingredients is extensive. According to the FDA report in January 2019, the volume of imported seafood, fresh fruit and vegetables are reaching 94%, 55% and 32%, respectively, in the USA (FDA, 2019b). The importing figure is showing a gradually increasing trend. There are a few reasons for such growth including decreased purchasing costs, limited local resources, and increased market demand (Ringsberg, 2014). However, global trade also increases the chances of food risks due to factors including geographical distance, cultural differences, the economy, and policies (Ghemawat, 2001). According to the sensitivity levels summarised by Ghemawat (2001), food products like meat and cereals, drinks and sugar, live animals and meat products are more sensitive to *cultural* distance, *administrative* distance, *geographic* distance, and *economic* distance, respectively.

Food quality and safety is a worldwide phenomenon and very much associated with the globalisation and outsourcing phenomenon (Yakovleva, 2007). More than just ingredients and raw materials, processes such as food processing and transportation are also involved (Mai et al., 2010). Safety issues not only contribute to the growing problem of public health issues but also have a significant impact on national economies. A few researchers have mentioned the potential negative impacts of the globalised food trade, such as the disparities that cause vulnerability in the food supply chain, inefficiency in communication, low transparency, low accountability, and lack of proper monitoring (Ghemawat, 2001; Roth et al., 2008; Yakovleva, 2007; Ringsberg, 2014; Mattevi & Jones, 2016). In 2018, 28 out of 382 food recalls in the USA were due to foreign material contamination (Maberry, 2019). The case example mentioned above, Sanlu milk scandal not only caused the death of babies in China but also contaminated pet food in the USA (Waldmeir, 2008). After the issue was exposed, many countries such as Japan and Singapore had to ban the import of any Chinese dairy products and recalled products which may contain suspicious ingredients (BBC, 2010; Huang, 2014).

Given the complexity of the food supply chain system, how to solve the related issues and be more sustainable is meaningful and timely. The following sections introduce the background information of the food supply chain, food traceability, and blockchain application. The food supply chain can provide a "holistic perspective", which allows us to identify the problem and propose possible solutions to food system issues by "understanding the interrelatedness of processes" (Yakovleva, 2007, p. 78; Kouhizadeh & Sarkis, 2018). Food traceability is a strategic approach for both preventing potential food hazards and helping with food recalls. By operating technological innovations, food traceability systems are becoming more efficient. Among all the technologies, blockchain is emerging and gaining more attention due to its uniqueness and huge potential. It is believed to bring a significant change to the food traceability systems.

1.2 Food Supply Chains

The food supply chain is considered

> a network of organizations, which through economic relations with each other enable the functioning of the supply chain for the production and distribution of food (Yakovleva, 2007, p. 76).

From this definition, the food supply chain is about physical products flow, information flow, and financial flows from "farm to fork" (Stevens, 1989; Dani & Deep, 2010). Folkerts and Koehorst (1997, p. 11) provided another definition:

> a food supply chain is defined as a set of interdependent companies that work closely together to manage the flow of goods and services along the value-added chain of

agricultural and food products, in order to realize superior customer value at the lowest possible costs.

In this case, the importance of collaboration and information flow, which refers to the transparency of the supply chain and the openness of discussions between stakeholders, are emphasised to achieve the supply chain objective, which is to satisfy customer demand at minimum costs (Stevens, 1989; Boehlje, 1999; Lindgreen & Hingley, 2003; Trienekens & Zuurbier, 2008; Dani & Deep, 2010; Feenstra et al., 2011). On average, from retailer to consumer, there are 27 intermediaries across three to five countries that process 240 copies of documents to each batch of the product, and 95% of the information from the documents has no value. Food tracking to the origin takes about seven to 14 days and requires several organisations to cooperate (Cointelegraph and VeChain, 2020). According to Folkerts and Koehorst (1997), the supply chain becomes more market-driven, which emphasises customer demand more. Consumers need food products to be healthy, safe, of good quality, and with various choices. Particularly after the COVID-19 pandemic, there is a rising awareness of food safety among consumers (Trienekens et al., 2012; Cointelegraph and VeChain, 2020; McKinnon, 2020).

1.2.1 Prevent Potential Food Hazards

There are two defences of food safety in the food system before and after food hazards. The first one includes policies and inspections to prevent potential contaminated food products flow in the supply chain, while the second defence is the food recall system when food hazards are identified (Karthikeyan & Garber, 2019). Food security is considered as a shared responsibility for the whole supply chain and all stakeholders, including producers, governments, and customers (Kaferstein, 2003; Trienekens et al., 2012; Storoy et al., 2013; Aung & Chang, 2014; Karthikeyan & Garber, 2019).

To maintain food quality and safety, governments and non-government organisations have contributed to propose certification schemes, such as legalise food laws, give indicators for food quality, and make regular checks, in order to "reach a defined performance and to make this known to stakeholders..." (Ringsberg, 2014, p. 567). The increasing global food business also means the increasing importance of global certification standards (Ringsberg, 2014). For example, the International Organization for Standardization (ISO) 22,005 Food Safety Standard required companies to launch the one-up (suppliers) and one-down (customers) principle to know their immediate suppliers and customers (ISO, 2007). The Sanitary and Phytosanitary Measures (SPS Agreement) was proposed in 1995 to ensure that exporters should always meet the quality and safety requirements in the import market (WTO, 1998). In the food law area, EC General Food law regulation 178/2002 requires a traceability system for food products in Europe (European Commission, 2002). For developed countries, for instance, American food and feed companies were required to register with the FDA to maintain records for traceability purposes (FDA, 2019c). Developing country governments also show a growing awareness of food quality and safety control due to globalisation (Trienekens & Zuurbier, 2008). For example, Chinese governments published more than 52 laws to address food hazards and increased the frequency of inspections (Tang et al., 2015).

For food companies, it is important to be proactive to identify potential supply chain risks and minimise the impacts (Dani & Deep, 2010). The food supply chain is considered as vast with multiple stakeholders and processes involved vertically and horizontally. However, it can be fragile in that one single unit failure can cause the collapse of the whole supply chain. For food products, extra attention is needed in the supply chain processes; factors such as temperature, humidity, and storage conditions can largely affect food quality and may cause safety issues (Aung & Chang, 2014). Hazard analysis at critical control points (HAPPC), therefore, has been developed to ensure the needed transparency and to prevent possible hazards (Aruoma, 2006; Trienekens & Zuurbier, 2008; Aung & Chang, 2014). This pre-check auditing method was recommended to many governments and policymakers to avoid excessive food testing (Lam et al., 2013). HAPPC principles include analysing hazards, identifying critical control points, building procedures to manage the control points, designing actions, verifying the whole system and record-keeping, and are used in many systems and regulations to ensure food quality and safety (Trienekens & Zuurbier, 2008). Besides running the risk analysing system, there are more things companies can do to prevent risks, such as giving hygiene education to employees (Kaferstein, 2003), building close and long-term relationship with suppliers (Lindgreen & Hingley, 2003), and complying with food safety policies (Kaferstein, 2003), among others.

Consumers also have responsibilities to ensure food safety (Kaferstein, 2003; Aruoma, 2006). A considerable number of illnesses and deaths are caused by unhygienic food preparation due to unhygienic cooking facilities, lack of personal hygiene, or cooking improperly, among other factors (Kaferstein, 2003). This situation is particularly serious in poverty areas, where the people have less awareness of cleanness, and insufficient infrastructures (Kaferstein, 2003). According to the WHO (2019), 35% of people from low- to middle-income countries lack water and soap for handwashing, while 19% of people have no improved sanitation. Although the number of diarrhoeal-related deaths has decreased in the past 25 years after improving water and sanitation, there were still about 1.3 million children killed by poor sanitation and unsafe food and water in 2015 (GBD Diarrhoeal Diseases Collaborators, 2015). Apart from development on consumer awareness and local infrastructures, Kaferstein (2003) also mentioned the lack of direct communication between consumers and other parties (governments and food companies) and suggested that close collaboration between partners is vital.

1.2.2 Food Recall

The second fence is food recall. Efficient food recall along supply chain is the last protection for customers (Karthikeyan & Garber, 2019). Food recall is a formal request made by companies to their customers based on the suspensions of dangers on food quality and safety (European Commission, 2002). In other words, food recall is the consequence of product failure. The food hazards mentioned above are only the tip of the iceberg. In 2018, there were 382 food products recalls in the USA, and the total food recalls are showing an increasing trend (Karthikeyan & Garber, 2019; Maberry, 2019). The issues can be caused by food contamination, excessive use of chemicals, inappropriate handling and storage, mislabelling, and undeclared allergens (Ringsberg, 2014; Maberry, 2019). For example, 160 out of 382 US food recalls in 2018 were caused by undeclared ingredients, such as milk or nuts, and 40% of the total recalls were due to microbiological contamination (Maberry, 2019).

During the products recall, it is not only the customers that face health threats; food companies also suffer economic loss from product recalls (Ringsberg, 2014). The average costs for one recall were $10 M for an American food company (Tyco Integrated Security, 2012). In 2011, 31 people were killed and thousands of people were infected by having German-produced bean sprouts (Sample, 2011). The outbreak brought huge compensation (210 m euros) to farmers and Germany suffered economic loss from restrictions on its exports of fresh food (BBC, 2011). In 2007, the American food brand ConAgra recalled 326 million pounds of Peter Pan and Walmart's peanut butter plus 99,953 cases of topping due to a salmonella outbreak (Nash,2007). The recall cost ConAgra company more than $78 m, plus 63% drop in sales in the year (Nash, 2007; Danovich, 2016).

Many researchers have mentioned that consumer confidence can drop significantly after recalls (Opara & Mazaud, 2001; Grunert, 2005; Kumar & Budin, 2006; Ringsberg, 2014; Tang et al., 2015). The large number of food hazards leads to increasing concerns over food quality and safety, which is also called "food scares" (Grunert, 2005). From the food safety survey in China in 2018, more than two-thirds (68.3%) of participants were concerned about food fraud problems (Fortune, 2018). For example, apart from Sanlu company, another 22 companies also involved in the milk scandal including other well-known national brands such as the Beijing Olympic dairy supplier, Yili company and one of the major national dairy brands of Mengniu (Huang, 2018). Moreover, other low-quality foods such as plastic seaweed and recycled food oil were also exposed and caused more panic on national food brands. Local authorities were also under investigation due to the suspension of bribery and covering up for criminals. All those scandals revealed the weakness of food law in China and destroyed the public trust in national food brands and government authorities (Huang, 2014). Grunert (2005) proposed that public perceptions of food risks do not apply in normal conditions, but can have a huge impact on customer confidence and buying choices during food hazards. Even 10 years after the Sanlu milk scandal, Chinese parents still tend to trust foreign brands more, and the fear of national food safety remains (Huang, 2018). Food recall-related indirect losses and further damages on brand reputation and sales are countless. Other examples of customer confidence loss caused sales drop include: the horse meat scandal mentioned above, caused the major retailer brand TESCO lost an estimate $408 m due to the international sales drop by 4.6% (Ringsberg, 2014); the whole egg industry lost about $100 m in revenue in 2010 due to the salmonella outbreak (Danovich, 2016).

1.3 Traceability

Once food recall happens, it is important to identify problems and take the appropriate actions during the "golden hours" (Dani & Deep, 2010). Slow reaction and delays can only cause more damage and costs (Roth et al., 2008). In other words, speed of response to the problem is crucial. In many cases, the recall tends to be very complicated and slow (Lindgreen & Hingley, 2003). For example, to issue a recall in the USA, a few steps are needed beforehand, such as conducting inspections, collecting samples, and clarifying risks (Chamlee, 2016). It is possible to take more than 100 days to complete all the steps. After the FDA was informed of the potential risk, the average recall for companies can take another 57 days (Mccallister, 2017). Moreover, even when the recall decision is finally issued, it is still hard for the FDA to know how completely the recall has been carried out (Danovich, 2016). That is, dangerous food can remain due to food recall failure (Mccallister, 2017; Karthikeyan & Garber, 2019). For example, I.M. Healthy soy nut butter spreads and granolas were recalled due to an E. Coli outbreak, which caused 32 illnesses in 2017 (Karthikeyan & Garber, 2019). However, the contaminated products were still sold in some places 6 months after recall was issued. This example shows an extreme example of inefficiency of food recall in the food industry, which can lead to deeper food safety concerns.

In this case, an efficient traceability system is considered as an important mechanism, "core enabler" or "effective corrective actions" to reduce the negative consequences (recall costs, products wastes, social impact, reputation damage) during recalls (Garcia-Torres et al., 2018; Mai et al., 2010; Mattevi & Jones, 2016; Moe, 1998; Opara, 2003, p. 103). Traceability was introduced in the 1990s in the fields of health, space, and military activities, and has attracted more attention in the food industry in recent decades due to the increasing number of food hazards (Ene, 2013). There are many definitions of traceability by different organisations which depend on the food industries. Traceability is formed by three main components: record-keeping, tracing, and tracking. Record-keeping is the essence of traceability; it collects information and allows the information to be retrieved when necessary (Manos & Manikas, 2010; Opara & Mazaud, 2001). Efficient recording is useful to isolate certain products or suppliers particularly when the food supply chain is vast. The more information, the quicker the recall can be (Moe, 1998). Tracing backward and tracking forward allow products information to be checked in any step within a supply chain, from the origins to customers—that is, "from farm to fork" (Aung & Chang, 2014; Pizzuti et al., 2014).

Moreover, it has the ability to provide a history and to localise selected products, by giving answers to these questions about food quality and safety, such as who (stakeholders), when(time), where (location), and why (causes) (Aung & Chang, 2014; Garcia-Torres et al., 2018). Therefore, traceability is not only a passive way to make food recalls but can also be used "in an active sense" to improve supply chain management and to gain consumer confidence (Jansen-Vullers et al., 2003; Moe, 1998, p. 403). Many researchers have discussed its ability to act as a proactive strategy/defence to monitor and prevent potential food quality and safety risks (Opara, 2003; Opara & Mazaud, 2001). It is worth noting that traceability itself cannot guarantee food safety and quality; rather, it can be used as tool that integrates with other quality assurance systems such as Hazard Analysis and Critical Control Point (HACCP), to assist in decreasing the number of potential incidents (Opara, 2003; Opara & Mazaud, 2001). In other words, traceability can give evidence of food products' quality meeting certain standards, by providing useful product information.

It is necessary to mention transparency when discussing traceability. Transparency is found to be closely related to consumer trust and confidence. Producers tend to provide information that benefits themselves to gain more profits, which can indirectly mislead customers and cause food crisis (Mao et al., 2018). Central authorities can also be tempted by bribes and cover up the truth. Many food incidents, such as the Sanlu milk scandal, were caused by lack of transparency in its supply chain at the first place and followed by the misuse of power by central authorities. Thus, customers are demanding to know more information before making their purchases, so more transparency of food supply chain is needed to keep customer confidence (Food Insights, 2019; Trienekens et al., 2012). One of the achievements in safeguarding food quality and safety is the compulsory labelling system introduced in the EU by the food law (European Commission, 2019; Wognum et al., 2011). The labelling rules provide a certain quality of products information such as ingredients and expiration dates and can help customers to make better choices before purchasing. By providing more transparency within the supply chain, parties will be able to have more knowledge of the current situations and make better decisions. The benefits include better inventory management, efficient transportation planning, and more accurate future demand forecasts among others.

1.4 Traceability Innovation

With the rapid advances in technology development, the wide use of computers, the internet and a variety of technological media has significantly changed the ways information is exchanged. Transactions and communication can be made via technological innovations to improve efficiency (Handfield & Bechtel, 2002).

Meanwhile, the costs of paperwork and labour can also be eliminated by digitising information.

Information systems are also more popular recently as information flow is crucial in food supply chains (Christopher, 2016; McMeekin et al., 2006). Companies tend to employ supply chain innovations to improve supply chain performance including improving customer services, reducing costs, or increasing efficiency (Franks, 2000; Hazen et al., 2012). Innovation is considered as something "new" which includes ideas, methods, or devices (Kahn, 2018). Supply chain innovation was defined by Arlbjorn et al. (2011, p. 8) as

> a change (incremental or radical) within the supply chain network, supply chain technology, or supply chain processes (or combinations of these) that can take place in a company function, within a company, in an industry or in a supply chain in order to enhance new value creation for the stakeholder.

Later, Storer et al. (2014, p. 490) defined it as follows:

> Supply chain innovation often involves collaborative and partnering relationships, particularly in terms of utilizing industry-led and industry-wide innovation, considered mutually beneficial, such as new technologies and information system

Canavari et al. (2010) proposed that technological limits can be one of the constraints that impact on the information flow in the supply chain. Christopher (2016) also pointed out that successful companies all have efficient information system to improve customer responsiveness. Many emerging technological innovations are employed in food supply chains to improve traceability. The innovations include Radio Frequency Identification (RFID), Wireless Sensor network (WSN), and Blockchain. The technologies are introduced fully in the next section, with an emphasis on blockchain applications.

1.5 Blockchain

Blockchain is a new concept that has attracted increasing attention in recent years. The global blockchain market in 2018 reached $583.5 million, and it is expected to reach $28 billion by 2025 (Meticulous Research, 2018). The key growth drivers are the increasing attention on blockchain, increasing adoption, and the growing trend of accepting cryptocurrency (Meticulous Research, 2018). The concept of blockchain was first announced by Satoshi Nakamoto in 2008. Nakamoto introduced a decentralised distributed database where everyone can access data and make transactions without central authorities. In this case, no single party in the blockchain can control or make changes to the whole database without total agreement from other users. Third parties such as governments and bank systems are not necessarily needed in transactions. The most successful practical example based on this concept is Bitcoin—a decentralised peer-to-peer electronic cash system. Financial use which includes transaction payments, cryptocurrency, and others accounts for the largest share of the global blockchain market according to Meticulous Research (2018). A

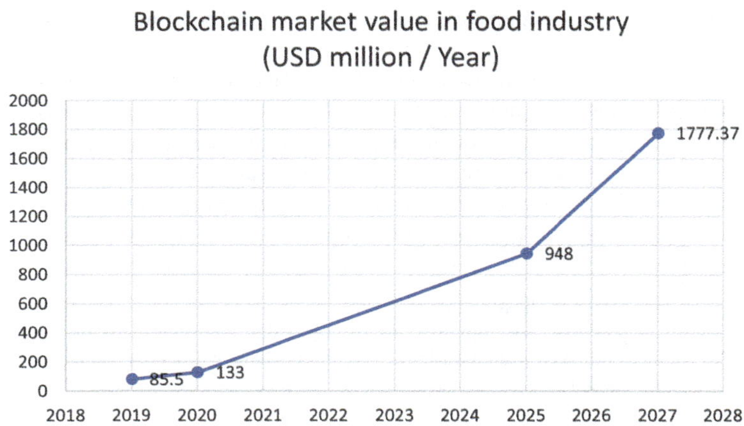

Fig. 1.1 Blockchain market value in the food industry (Source: Research and Markets, 2020; Verified Market Research, 2020)

significant body of research is also focused on its financial uses. Chen et al. (2017a, b) proposed a blockchain-based digital wallet to make transactions. Folkinshteyn and Lennon (2016) used the technology acceptance model (TAM) to examine the adoption intentions of the blockchain-based financial platform. Apart from its financial use, the blockchain concept has now been widely discussed and investigated by many fields where trust and value are important, such as banking systems, property management, retailing, diamonds supply chain, pharmaceutical industry, sustainability (Kouhizadeh & Sarkis, 2018) as well as food supply chains (Chen et al. 2017a, b; Lin et al., 2017; Sylim et al., 2018; Zhao et al., 2016).

For food supply chains, blockchain is adopted to adjust the information inequality between supply chain stakeholders. Blockchain is going to improve information transparency and improve traceability system (Petersen et al., 2018). According to a joint report by Cointelegraph and VeChain (2020), the combination of blockchain and IOT solution can save $70 billion in costs and create $47 billion income in the global food industry. The market value of blockchain in the food industry reached $85.5 million in 2019, $133 million in 2020, and is predicted to reach about $948 million in 2025 (Research and Markets, 2020) and $1777.37 million by 2027 (Verified Market Research, 2020). By 2023, 10% of the food products will be tracked by blockchain, and $300 billion worth food products will be blockchain-traceable by 2027 (Cointelegraph and VeChain, 2020). It is a promising solver for the current trust issues and will bring back public confidence in the food industry. It is one of the promising technologies that is expected to make revolutionary changes to supply chain management (Kouhizadeh & Sarkis, 2018). There is an increasing trend of research papers of blockchain applications in food supply chain management. All the previous papers are analysed by the content-based analysing method in the following chapter (Fig. 1.1).

1.5.1 Blockchain Applications

As blockchain is expected to improve traceability in food supply chains by providing transparency of information flow, trust between stakeholders, efficiency, and speed of tracing certain products it is gaining more attention for its promising potential to address supply chain issues. Many companies and organisations are engaged to apply blockchain to improve supply chain efficiency and products' quality such as IBM, Walmart, Moyee Coffee, and WWF (as summarised in Table 1.1).

IBM, after 18 months of testing, launched its blockchain-based food traceability platform—IBM food Trust platform, which allows the engagement of various stakeholders in the ecosystem including retailers, wholesalers, and suppliers (Stanley, 2018). The first adopters of the IBM Food Trust platform are French retailer *Carrefour*, a well-known food producer; *Nestle,* a supplier, and *BeefChain*. As a food traceability platform, it has three modules: tracing module, certification module, and data entry and access module. The *tracing* module can track products within

Table 1.1 The application of blockchain in food supply chain

Industry	Companies	Achievements
Food industry	IBM—IBM Food Trust Platform	1. Products' tracking 2. Certifying products
Food retailing industry	Walmart & IBM & Tstinghua university—Walmart food safety collaboration centre	1. Improving traceability system 2. Improving transparency 3. Assuring products authenticity and safety
Coffee industry	Moyee Coffee & Bext360 & KripC—KripC platform	1. Improving transparency 2. Eliminating middlemen 3. Cost reduction
Seafood industry	WWF & ConsenSys & Traseable	1. Eliminating illegal-caught products and unethical labour 2. Improve transparency
Seafood industry	Hyperledger—Sawtooth	1. Combining sensors and blockchain to track products. 2. Improving supply chain transparency and accountability
Agriculture products	AgriDigital commodity management platform—AgriDigital & CBH	1. Matching title transfer of the grain asset to payment 2. Supply chain provenance and traceability
Agri-Food &seafood	OwlChain & AMIS blockchain structure	1. Creating an open and tamper-resistant food provenance system which provide customer with more buying confidence, and help farmers to earn more 2. Helping on the products label authenticity and reduce unsustainable fishing behaviours.
Farming	ripe.io	1. Creating transparency, trust, and honesty in food supply chain

the supply chain, even across borders. The *certification* module is about nongovernment organisations certifying the products, such as fair trade or organic. *The data entry and access* module allows stakeholders such as growers to upload and manage the products' data. The latter is free to all stakeholders, while the first two modules cost a little every month. The data entry and access module is also simple to use and does not require users to be experts.

The world's largest retailer Walmart has also been working with IBM and Tsinghua University to create a blockchain-based food security system that targets traceability and food fraud (SupplyChain247, 2016). According to Yiannas (2018), in a pilot study, the time of tracing mangoes reduced from nearly 7 days to 2.2 s by using a blockchain system. By building a Walmart food safety collaboration centre in China and tracking pork, customers can now obtain maximum information such as farm details, factories, processing data, storage data, expiration dates, and shipping details on the products they purchase. The in-depth information potentially proves products' authenticity and addresses food safety issues. Due to the great success of the pilot studies, in September 2019, Walmart announced that it is compulsory for their leafy green vegetable suppliers to use blockchain to trace products (Kharif, 2018). After launching the plan, over 100 of its suppliers will have to use blockchain technology in 2019. Walmart also hoped to implement a blockchain-based traceability system to other fresh products within 2019. Walmart's vice president of food safety Frank Yiannas asserted: "This is a smart, technology-supported move that will greatly benefit our customers and transform the food system, benefitting all stakeholders". Blockchain engagement can also benefit other stakeholders, for example, retailers can manage products' shelf-life according to records (Yiannas, 2018).

Moyee Coffee is working with Bext360 and KripC on the world's first coffee blockchain project to improve coffee transparency and improve fairer trade (Moyee Coffee Ireland, 2017). Similar to IBM Food Trust platform, KripC platform also allows all stakeholders including farmers, roasters, retailers, and consumers to access data. As a complex industry, coffee producers used to only receive 2% added value as there were too many middlemen in the supply chain. The platform makes the whole coffee supply chain more transparent and reduces unnecessary costs, such as omitting the middlemen. Therefore, the platform not only provides customers with data and sources of the coffee they drink but also benefits the lower tier of the supply chain with more value added. The paperwork and physical inspectors can also be reduced by the blockchain platform and save up to 0.80 euros per pound of coffee (Moyee Coffee Ireland, 2017). The entire blockchain coffee supply chain was launched in 2017 and proved the traceability of 60,000 kg of coffee by June 2018 (Best360, 2018).

In terms of the seafood supply chain, the world-known environmental organisation, the World Wildlife Fund (WWF), is working with some partners such as technology companies ConsenSys and Traseable to develop a blockchain-based system in order to improve the traceability in the seafood supply chain (WWF, 2018). This innovative project can improve the transparency in the seafood supply chain, this means that customers are allowed to know the stories by scan packing behind the seafood they bought such as origins, vessels, and fisherman. Due to the accessible information, therefore, unsustainable seafood such as illegal-caught products or unethical labour in the seafood industry can be avoided. Taiwan-based OwlChain platform does a similar job, which protects seafood label authenticity and improves fishing sustainability (OwlTing, 2017). Another blockchain application, Hyperledger Sawtooth, is being used in the seafood supply chain industry to improve its traceability and accountability (Sawtooth, 2018). It is a combination of sensors and blockchain to trace sea food products. From the sea to the table, products are recorded for their whole journey, which includes the location, temperature, humidity, motion, shock, tilt, and ownership transfers (Sawtooth, 2018). The recorded information allows customers to access before making their purchases. Customers, therefore, are guaranteed the products' quality and authenticity by blockchain and the application also benefits the local fishing industry.

For the agri-food industry, several start-ups have created blockchain structures to ensure food safety and quality, such as AgriDigital and CBH, OwlChain and AMIS, and ripe.io. In 2017, Australia-based AgriDigital and CBH launched a pilot study to test blockchain in the grain industry and achieved satisfactory results (AgriDigital and CBH Group, 2017; Antonovici, 2017). It is not only the growers and retailers who benefit from the instant payments, customers can also have buying confidence due to the efficient products' traceability. Based on Ethereum (Smart contract blockchain platform) platform, OwlChain and its partner AMIS created a Taiwan-based Agri-Food trade platform (OwlTing, 2017). By using blockchain, consumers gain more visibility on and trust in food quality and safety and are willing to pay higher prices. In this case, by earning more and paying low commission fees to the platform, farmers can have a higher income. Ripe.io is a platform that combines blockchain and IoTs and is aiming to provide transparency and trust between stakeholders including farmers, distributors, and consumers.

Table 1.2 provides more applications of blockchain in various industries. WaBi and VeChain are developed to prove products' authenticities including food, wine, cosmetics, and agricultural goods, among others. To eliminate fraud products, blockchain is popular in the pharmaceutical and luxury goods industries. BlockRx and Modumf are used on drugs delivery, to control temperatures and monitor processes. Everledger developed the Time-Lapse Protocol to prevent "blood diamonds" by tracing diamonds origins. There are also various blockchain applications in global shipping.

1.6 Research Objectives

The following sections present the research questions that this research aims to answer, and the research structure.

Table 1.2 Other examples of blockchain applications

Blockchain ledger	Area	Achievement	Details
WaBi (WABI)	Food industry (liquor, baby food), cosmetic industry.	Products authenticity	Inspired by Sanlu milk scandal, WaBi is a combination of RFID chips, QR codes, and blockchain, which allows customer to scan label before making purchase (Quittem, 2018).
VeChain platform	Agricultural, wine, luxury goods	Cost efficiency, and security, products authenticity and information transparency	It is a decentralised platform for business solutions without any intermediary. Stakeholders can share and manage products' information by using blockchain. For example, manufacturing can provide products' data to ensure products' authenticity and quality (Benson, 2018; Zwanenburg, 2018).
BlockRx (iSolve & Swiss pharmaceutic company)	Pharmaceutical industry	1. Reducing counterfeit drugs 2. temperature control on certain drugs	1. As blockchain can improve the transparency of the supply chain, stakeholders of the supply chain can make sure of the authenticity of the source of drugs. 2. Decentralised characteristic of blockchain technology can also reduce the risk of technology hackers and irresponsible authority. 3. Blockchain benefits the pharmaceutical cold chain. Blockchain can keep the temperature information transparent along the supply route and remind the manager or driver to monitor it on time.
Modum Blockchain (Swiss post & Modrum)	pharmaceutical	Temperature monitoring on delivery products	It is using smart contracts combined with IoT technologies to delivery temperature-sensitive products. The temperature data will be continuously recorded and automatically send notifications to senders and receivers if the temperature is out of limits. This technology can provide

(continued)

Table 1.2 (continued)

Blockchain ledger	Area	Achievement	Details
			customers with quality products and provide transparency during delivery (Das, 2018).
Ambrosus (AMB)	Food & Pharma industry	Supply chain optimisation, logistic tracking, quality assurance, anti-counterfeiting	Combination of blockchain and IoT (RFID chips, sensors, and QR codes) to track movement and temperature to achieve smarter, healthier and transparent food and pharma ecosystem connections with their stakeholders (Gutteridge, 2018; Quittem, 2018).
Time-lapse protocol (Everledger)	Diamonds	1. Eliminating counterfeit products 2. Eliminating "Blood diamonds"	For high-value luxury goods such as diamonds, blockchain is also helpful to reduce risks of counterfeit products and unethical products. 1. Blockchain-based Diamond Time-lapse Protocol is able to protect the transparency of the diamond trade, to assure the authenticity of diamonds and to provide the provenance of diamonds (Diamond Time-Lapse, 2018). 2. The technology can trace diamonds from the origin and eliminate the "blood diamonds"—diamonds that are produced to support unethical activities such as wars (Sunny, 2018). From mining to retailing, every process of the diamond supply chain is verified and can be checked by the public.
Blockfreight	Global cargo shipping industry	Improving market access Building financial strength in global trade	1. Smart contracts to securely, permanently define the bill of lading, terms of payment, and other key elements to a completed cargo shipment, built on the Ethereum blockchain. 2. A tradeable token built as a Counterparty asset, which

(continued)

Table 1.2 (continued)

Blockchain ledger	Area	Achievement	Details
			is secured by the bitcoin blockchain. This token pays for the transaction fees and effectively eliminates spam on the decentralised system. 3. Storage of the bill of lading and other documents too large to fit into a bitcoin block using the IPFS (Inter Planetary File System) protocol (Coleman, 2016).
ShipChain platform (ShipChain)	Shipping	Creating transparency and efficiency when within supply chain and during tracing and tracking items	It is a fully integrated system across the whole supply chain to provide transparency and efficiency. From leaving the factories to the harbours, from retailers to customers, all products can be traced and tracked. By using smart contract, shippers and carriers will have more visibility in the supply chain (ShipChain, 2018).
TradeLens (Maersk and IBM)	Global supply chain	Providing a shipping information pipeline paperless trade (Maxie, 2018).	It is an end-to-end blockchain-based platform to improve efficiency, provide transparency and trust, and reduce the risks in global supply chains (Biazetti, 2018; Maxie, 2018). At least 90 companies have joined and share data on the platform (Tradelens, 2018).
Waltonchain	Supply chain	Complete data sharing to reach information transparency, eliminating counterfeited products	It combines both RFID chips and blockchain to achieve Value internet of Things, which is a decentralised ecosystem of interconnectivity (Brauer, 2017). It aims to create a genuine, believable, traceable business model with total shared data and transparent information, and to address trust issues within the supply chain, and provides customers with authentic and good quality of products (Waltonchain, 2018).

(continued)

Table 1.2 (continued)

Blockchain ledger	Area	Achievement	Details
Ethereum	Supply chain	Reduce third parties' interferences, improve efficiency	It is a blockchain platform that runs smart contracts without any interference. By running smart contract on blockchain, programs will be operated automatically, which can eliminate third parties. Developers can use the Ethereum platform to create any operations (Blockgeeks, 2018).

1.6.1 Research Questions

The above sections reviewed the food supply chain, food traceability, technology innovations on food traceability and, particularly, the application of blockchain in the food supply chain. A few identified research gaps are addressed in this paper. First, more academic research is needed on the application of blockchain in food supply chain management although, according to recent trends, blockchain is becoming very popular in supply chain management and some pioneering studies have discussed similar topics. Tian (2017) introduced the blockchain concept in the food supply chain by building a framework that combines IoT technologies. Kshetri (2018) and Verhoeven et al. (2018) also discussed how companies adopted blockchain applications in the food industry by applying a case study method. A limited number of articles have discussed t the application of blockchain in food supply chain management specifically in how the pioneer companies have implemented blockchain to improve food quality, enhance traceability, and eliminate food incidents. Secondly, blockchain is a very new concept which was only proposed a decade ago and most of the industry applications only started recently. Among the previous research, most focused on the theoretical concepts rather than on empirical evidence while the reality can be far complicated than the assumptions Third, Tables 1.1 and 1.2 list many companies with the application of blockchain in various industries. The practice seems to be leading the research, while the latter needs to catch up to provide guidance on the practice.

Therefore, this research focuses on the application of blockchain in food supply chain management. To provide deeper insights into the practical uses of blockchain and its influences in the food supply chain, this research adopts a multiple case study method to investigate the following two questions:

Research Question 1): Why do companies apply blockchain in food supply chain management and based on what threats or opportunities in the food supply chain?

Research Question 2): How has blockchain been applied in food supply chain and how can it influence the food industry?

More specifically, food is one of the basic human needs which requires safety and quality. Due to the development of global trade, foodborne diseases happen more often nowadays. The incidents not only harm people's health and ruin public trust in the food market but also can impose pressure on national medical systems and national exports businesses (Tian, 2016). The increasing number of food safety incidents lead to the customers' increasing demand for food quality and more informed knowledge about the food-producing processes. The incidents also encourage companies to adopt better innovative approaches to eliminate food risks.

The food supply chain can be improved to eliminate food insecurity risk, reduce food recall costs, and achieve sustainability. To retain food products' freshness and quality, efficient information systems that enable efficient information flows are highly required within food supply chain. Blockchain is one of the most promising technologies that is expected to bring revolutionary changes in the food industry (Kouhizadeh & Sarkis, 2018). It is expected to provide transparency and efficiency, to address public trust and confidence, and to improve traceability within food supply chains (Kouhizadeh & Sarkis, 2018; Saberi et al., 2018). Customers have the rights to know the stories behind the products they have purchased, while companies can also improve supply chain efficiency and achieve operation excellence. As there is no central power over the supply chain, it also prevents misleading or intervention by third parties and avoids the potential of bribery (Tian, 2017).

This research adopts the innovation process model and practice-based view (PBV) to answer the research questions. The innovation process model can help to explain the process of innovative thinking and answer the first research question: *Why do companies apply blockchain in food supply chain management and based on which threats or opportunities in the food supply chain*? Moreover, the model also presents the activities that the companies adopt to put the idea in practice. Therefore, it can also help to answer the second research question: *How has blockchain been applied in food supply chain and how can it influence the food industry?*

The PBV is adopted in this research for two reasons. Firstly, blockchain application is in its early stages. Only a limited number of food companies have adopted blockchain. Most literature on the topic is also theory-based. In this case, learning from practice is extremely useful as it can provide a practical view of blockchain implementation. The PBV can help to find the key elements to success and the common barriers during blockchain implementation. Second, the PBV can perfectly combine with the innovation process model to provide a more comprehensive view and a complete picture of blockchain implementation. Ellstrom (2010) suggested that there are two kinds of innovation: *potential* innovation, and *actual* innovation. Potential innovation has not yet shown the benefits; however, the actual innovation is something which, "rather than a creative idea, invention or discovery, can only be determined on the basis of its practical application and use" (Ellstrom, 2010, p. 3). In the research by Su et al. (2011), an innovation performance was added after the innovation process model. However, the research has only studied one hospital, which is hardly to apply PBV to the performance. Therefore, this research adopts the PBV to systematically provide the business performance that follows the innovation process model. The innovation process model answers the research questions of "why" and "how". PBV can summarise the common practices and present the business performance from blockchain implementation.

This research contributes to the following aspects: *first*, it presents the issues within food supply chains that can threaten food quality and safety. In other words, the main drivers for simulating food companies to adopt blockchain are discussed here. Therefore, this paper investigates how food companies react to the threats and tackle the problems by using blockchain.

Second, this research investigates how food companies engage with blockchain applications in the reality. As blockchain is a relatively new technology and still in its early stages, there are very few practical blockchain applications within the food industry. As far as this research is concerned, the earliest blockchain platform—IBM food trust was operated in 2016, and the earliest published paper about the blockchain-based food supply chain was presented by Tian (2016). The blockchain theory was proposed a decade ago, while the food supply chain-focused blockchain only started to be discussed 5 years ago. The weak knowledge and experience foundation requires more research in this field. This research, therefore, aims to provide a better understanding of blockchain and its potentials. By using multiple case studies, the research also provides a clearer view of the benefits of applying blockchain in the food industry.

Third, most of the related research is conceptual and has provided positive outcomes. It is reasonable to suspect that most papers tend to come out with positive outcomes to follow the blockchain "hype" and to meet public expectations. This research, therefore, also investigates the challenges of adopting blockchain applications, and how companies can solve the problems. Overall, the research aims to provide a clearer and empirical view of blockchain application in the food industry and give guidance to future researchers. This research provides both theoretical and practical implications.

1.6.2 Research Structure

The research structure is as follows: Chap. 2 provides a systematic literature review on blockchain and food supply chain management. The methodology is presented in Chap. 3 which presents case study research method. Three within-case analyses are presented in Chap. 4 (BeefLedger), Chap. 5 (W company) and Chap. 6 (FairChain), respectively. The cross-case analysis findings based on case studies are presented in Chap. 7. Chapter 8 is a discussion chapter that compares the observations with the existing literature. Chapter 9 is the summary and conclusion of the research and presents the implications and contributions (Fig. 1.2).

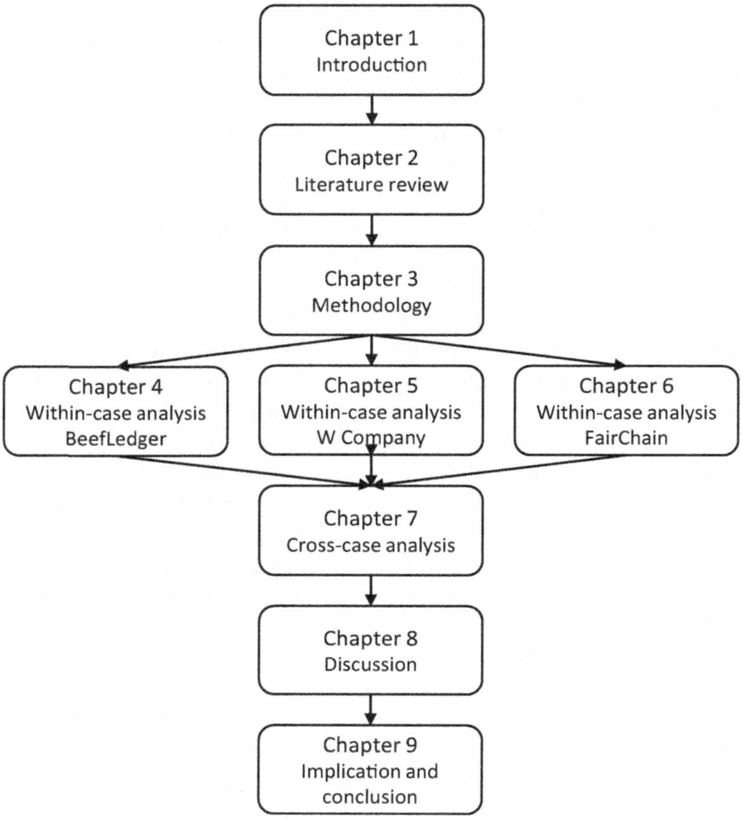

Fig. 1.2 Research structure

References

AgriDigital and CBH Group. (2017). *Solving for supply chain inefficiencies and risks with blockchain in agriculture*. Accessed May 02, 2019, from https://daks2k3a4ib2z.cloudfront.net/593ba04f0052061059d5383e/5a0be176b08b700001502550_AgriDigital%20and%20CBH%20Blockchain%20Pilot%20Report.pdf

Antonovici, A. (2017). *Australia's CBH Group, AgriDigital successfully test blockchain in grain industry*. Cryptovest. Accessed October 25, 2018, from https://cryptovest.com/news/australias-cbh-group-agridigital-successfully-test-blockchain-in-grain-industry/

Arlbjorn, J. S., de Haas, H., & Munksgaard, K. B. (2011). Exploring supply chain innovation. *Logistics Research, 3*(1), 3–18.

Aruoma, O. I. (2006). The impact of food regulation on the food supply chain. *Toxicology, 221*(1), 119–127.

Aung, M., & Chang, Y. (2014). Traceability in a food supply chain: Safety and quality perspectives. *Food Control, 39*, 172–184.

BBC. (2010). *Chinese dairy executives charged in tainted milk case*. Accessed October 25, 2018, from http://news.bbc.co.uk/1/hi/world/asia-pacific/8443437.stm

BBC. (2011). *Deaths from E. coli still rising in Germany*. Accessed October 30, 2018, from https://www.bbc.co.uk/news/world-europe-13746682

Benson, C. (2018). *VeChain Overview – Main features and partnerships*. Accessed November 18, 2018, from https://boxmining.com/VeChain/

Best360. (2018). *Best360 powers blockchain traceability for Token, the first community-owned coffee brand from Moyee Coffee and FairChain Foundation*. GlobeNewswire. Accessed April 28, 2019, from https://www.globenewswire.com/news-release/2018/06/22/1528470/0/en/Bext360-Powers-Blockchain-Traceability-For-Token-The-First-Community-Owned-Coffee-Brand-From-Moyee-Coffee-and-FairChain-Foundation.html

Biazetti, A.(2018). *What is freight fluidity and how can blockchain help?* Medium. Accessed October 30, 2018, from https://medium.com/tradelens-blog/what-is-freight-fluidity-and-how-can-blockchain-help-29681f7077f9

Blockgeeks. (2018). *What is ethereum? The most comprehensive guide ever!* Blockgeeks. Accessed October 30, 2018, from https://blockgeeks.com/guides/what-is-ethereum/

Boehlje, M. (1999). Structural changes in the agricultural industries: How do we measure, analyse and understand them? *American Journal of Agricultural Economics, 81*(5), 1028–1041.

Brauer, E. (2017). *Woltonchain: An interconnection between the physical and virtual ecosystems*. Invest In Blockchain. Available at: https://www.investinblockchain.com/waltonchain/. Accessed on: 15.02.2021.

Business & Human Rights Resource Centre. (2010). *Sanlu lawsuits (re tainted milk in China)*. Business-humanrights.org. Accessed February 15, 2021, from https://www.business-humanrights.org/en/latest-news/sanlu-lawsuits-re-tainted-milk-in-china/

Canavari, M., Centonze, R., Hingley, M., & Spadoni, R. (2010). Traceability as part of competitive strategy in the fruit supply chain. *British Food Journal, 112*(2), 171–186.

Chamlee, V. (2016). *Why does the FDA take so long to recall food?* EATER. Accessed November 25, 2018, from https://www.eater.com/2016/8/1/12343626/fda-food-recalls-chuck-schumer

Chen, P. W., Jiang, B. S., & Wang, C. H. (2017a). Blockchain-based payment collection supervision system using pervasive bitcoin digital wallet. In *2017 IEEE 13 International Conference on Wireless and Mobile Computing, Networking and Communications (WiMob)*, pp. 139–146.

Chen, S., Shi, R., & Ren, Z. (2017b). A Blockchain-based supply chain quality management framework. In *The fourteenth IEEE International Conference on E-Business Engineering*, pp. 172–177.

Christopher, M. (2016). *Logistics & supply chain management* (5th ed.). Pearson UK.

Christopher, M., Mena, C., Khan, O., & Yurt, O. (2011). Approaches to managing global sourcing risk. *Supply Chain Management, 16*(2), 67–81.

Cointelegraph and VeChain. (2020). *When blockchain meets IoT: Ensuring food safety in the 2020s*. Cointelegraph. Accessed February 02, 2021, from https://s3.cointelegraph.com/storage/uploads/view/4612f245e3cf8bc227084dff049f4016.pdf

Coleman, L. (2016). *Blockfreight taps Blockchain technology to disrupt global shipping*. CCN. Accessed October 30, 2018, from https://www.ccn.com/blockfreight-taps-blockchain-technology-disrupt-global-shipping/

Dani, S., & Deep, A. (2010). Fragile food supply chains: Reacting to risks. *International Journal of Logistics: Research and Applications, 13*(5), 395–410.

Danovich, T. (2016). *How food recalls really work: Companies pull 8,000 products off the shelves each year. Where do they go- and how much does that cost?* Eater. Accessed April 12, 2019, from https://www.eater.com/2016/1/26/10830614/food-recalls-how-they-work

Das, S. (2018). *Swiss post taps Modum Blockchain for temperature-monitoring of shipments*. CCN. Accessed October 30, 2018, from https://www.ccn.com/swiss-post-taps-modum-blockchain-for-temperature-monitoring-of-shipments/

Diamond Time-Lapse. (2018). *Do you know your diamonds?* Everledger. Accessed October 27, 2018, from https://diamonds.everledger.io

Ellstrom, P. E. (2010). Practice-based innovation: A learning perspective. *Journal of Workplace Learning, 22*(1/2), 27–40.

Ene, C. (2013). The relevance of traceability in the food chain. *Economics of Agriculture, 2*(60), 287–297.

European Commission. (2002). Council Regulation (EC) No 178/2002 of the European parliament and of the council. *Official Journal of the European Union, L31*, 1–24.

European Commission. (2019). *Food Law general requirements*. Accessed June 02, 2019, from https://ec.europa.eu/food/safety/general_food_law/general_requirements_en

FDA. (2019a). *Food safety: Importance for at-risk groups*. FDA. Accessed December 12, 2018, from https://www.fda.gov/food/people-risk-foodborne-illness/food-safety-importance-risk-groups

FDA. (2019b). *FDA strategy for the safety of imported food*. FDA. Accessed December 12, 2018, from https://www.fda.gov/food/importing-food-products-united-states/fda-strategy-safety-imported-food

FDA. (2019c). *Registration of food facilities and other submissions*. FDA. Accessed December 12, 2018, from https://www.fda.gov/food/guidance-regulation-food-and-dietary-supplements/registration-food-facilities-and-other-submissions

Feenstra, G., Allen, P., Hardesty, S., Ohmart, J., & Perez, J. (2011). Using a supply chain analysis to assess the sustainability of farm-to-institution programs. *Journal of Agriculture, Food System, and Community Development, 1*(4), 69–85.

Folkerts, H., & Koehorst, H. (1997). Challenges in international food supply chains: Vertical co-ordination in the European agribusiness and food industries. *Supply Chain Management: An International Journal, 2*(1), 11–14.

Folkinshteyn, M., & Lennon, M. (2016). Braving bitcoin: A technology acceptance model (TAM) analysis. *Journal of Information Technology Case and Application Research, 18*(4), 220–249.

Food Insights. (2019). *2019 Food & Health Survey*. Food Insights Org. Available at: https://foodinsight.org/wp-content/uploads/2019/05/IFIC-Foundation-2019-Food-and-Health-Report-FINAL.pdf. Accessed on: 24.03.2019.

Fortune, A. (2018). *Food safety survey 2018: The results*. Food Manufacture. Accessed March 24, 2019, from https://www.foodmanufacture.co.uk/Article/2018/07/05/Food-safety-survey-results

Franks, J. (2000). Supply chain innovation. *Work Study, 49*(4), 152–155.

Garcia-Torres, S., Albareda, L., Garcia, M. R., & Seuring, S. (2018). Traceability for sustainability-literature review and conceptual framework. *Supply Chain Management: An International Journal, 24*(1), 85–106.

GBD Diarrhoeal Diseases Collaborators. (2015). Estimates of global, regional, and national morbidity, mortality, and aetiologies of diarrhoeal diseases: A systematic analysis for the Global Burden of Disease Study 2015. *The Lancet Infectious Diseases*. Accessed April 12, 2019, from https://www.thelancet.com/action/showPdf?pii=S1473-3099%2817%2930276-1

Ghemawat, P. (2001). Distance still matters, the hard reality of global expansion. *The Harvard Business Review*. Accessed November 14, 2018, from https://hbr.org/2001/09/distance-still-matters-the-hard-reality-of-global-expansion

Grunert, K. G. (2005). Food quality and safety: Consumer perception and demand. *European Review of Agricultural Economics, 32*(3), 369–391.

Gutteridge, D. (2018). *What is Ambrosus (AMB)?* Invest in Blockchain. Accessed November 14, 2018, from https://www.investinblockchain.com/what-is-ambrosus-amb/

Handfield, R. B., & Bechtel, C. (2002). The role of trust and relationship structure in improving supply chain responsiveness. *Industrial Marketing Management, 31*(4), 367–382.

Hazen, B. T., Overstree, R. E., & Cegielski, C. G. (2012). Supply chain innovation diffusion: Going beyond adoption. *The International Journal of logistics Management, 23*(1), 119–134.

Huang, Y. Z. (2014). The 2008 Milk scandal revisited. *Forbes*. Accessed October 25, 2018, from https://www.forbes.com/sites/yanzhonghuang/2014/07/16/the-2008-milk-scandal-revisited/#4f817e674105

Huang, E. (2018). *Ten years after China's infant milk tragedy, parents still won't trust their babies to local formula*. QUARTZ. Accessed October 25, 2018, from https://qz.com/1323471/ten-years-after-chinas-melamine-laced-infant-milk-tragedy-deep-distrust-remains/
ISO. (2007). *ISO 22005:2007: New ISO standard to facilitate traceability in food supply chain*. ISO. Accessed December 15, 2018, from https://www.iso.org/standard/36297.html
Jansen-Vullers, M. H., van Dorp, C. A. V., & Beulens, A. J. M. (2003). Managing traceability information in manufacture. *International Journal of Information Management, 23*(5), 395–413.
Kaferstein, F. (2003). Foodborn diseases in developing countries: Aetiology, epidemiology and strategies for prevention. *International Journal of Environmental Health Research, 13*(1), 161–168.
Kahn, K. B. (2018). Understanding innovation. *Business Horizons, 61*(3), 453–460.
Karthikeyan, V., & Garber, A. (2019). *How safe is out food: Recent trends and case studies, and what they mean for out health*. U.S. PIRG Education Fund. Accessed December 12, 2018, from https://uspirg.org/sites/pirg/files/reports/WEB_USP_Safe-Food-Report_Jan19.pdf
Kharif, O. (2018). Walmart tells leafy-green suppliers to start using blockchain. *The Seattle Times*. Accessed October 19, 2018, from https://www.seattletimes.com/business/walmart-tells-leafy-green-suppliers-to-start-using-blockchain/
Kouhizadeh, M., & Sarkis, J. (2018). Blockchain practices, potentials, and perspectives in greening supply chains. *Sustainability, 10*(10), 1–16.
Kshetri, N. (2018). 1 Blockchain's roles in meeting key supply chain management objectives. *International Journal of Information Management, 39*, 80–89.
Kumar, S., & Budin, E. M. (2006). Prevention and management of product recalls in the processed food industry: A case study based on an exporter's perspective. *Technovation, 25*(56), 739–750.
Lam, H., Remais, J., Fung, M. C., Xu, L., & Sun, S. S. (2013). Food supply and food safety issues in China. *The Lancet, 381*(9882), 2044–2053.
Lin, I. C., Shin, H., Liu, J. C., & Jie, Y. X. (2017). Food traceability system using blockchain. In *Proceedings of the 79th IASTEM International Conference, Tokyo, Japan*, pp. 59–64.
Lindgreen, A., & Hingley, M. (2003). The impact of food safety and animal welfare policies on supply chain management – The case of the Tesco meat supply chain. *British Food Journal, 105*(6), 328–349.
Maberry, T. (2019). A look back at 2018 food recalls. *Food Safety Magazine*. Accessed April 13, 2019, from https://www.foodsafetymagazine.com/enewsletter/a-look-back-at-2018-food-recalls-outbreaks/
Mai, N., Bogason, S. G., Arason, S., Arnason, S. V., & Matthiasson, T. G. (2010). Benefits of traceability in fish supply chains – Case studies. *British Food Journal, 112*(9), 976–1002.
Manos, B., & Manikas, I. (2010). Traceability in the Greek fresh produce sector: Drivers and constraints. *British Food Journal, 112*(6), 640–652.
Mao, D., Wang, F., Hao, Z., & Li, H. (2018). Credit evaluation system based on Blockchain for multiple stakeholders in the food supply chain. *International Journal of Environmental Research and Public Health, 15*(8), 1627.
Mattevi, M., & Jones, J. A. (2016). Traceability in the food supply chain: Awareness and attitudes of UK small and medium-sized enterprises. *Food Control, 64*, 120–127.
Maxie, E. (2018). *Top Blockchain use cases for supply chain management*. VeryTechnology. Accessed October 22, 2018, from https://www.verypossible.com/blog/top-blockchain-use-cases-for-supply-chain-management
Mccallister, D. (2017). *FDA takes too long to recall tainted food, Federal investigator says*. NPR. Accessed April 12, 2019, from https://www.npr.org/sections/thetwo-way/2017/12/28/573987023/fda-takes-too-long-to-recall-tainted-food-federal-investigators-say
McKinnon, T. (2020). *The top 8 retail trends in China to watch*. Indigo9Digital. Accessed February 03, 2021, from https://www.indigo9digital.com/blog/top-shopping-trends-from-china

McMeekin, T. A., Baranyi, J., Bowman, J., Dalgaard, P., Kirk, M., Ross, T., Schmid, S., & Zwietering, M. H. (2006). Information systems in food safety management. *International Journal of Food Microbiology, 112*(3), 181–194.

Meticulous Research. (2018). *Blockchain technology market by industry vertical (BFSI, Government, Healthcare, Retail and E-commerce, Transportation and Logistics) application (payments, exchanges, smart contracts documentation, digital identity) and geography - Global Forecast to 2025*. Information & Communication Technology. Accessed June 26, 2019, from https://www.meticulousresearch.com/product/blockchain-market/

Moe, T. (1998). Perspectives on traceability in food manufacture. *Trends in Food Science & Technology, 9*(5), 211–214.

Moyee Coffee Ireland. (2017). *World's First Blockchain Coffee Project*. Medium. Accessed October 19, 2019, from https://medium.com/@MoyeeCoffeeIRL/worlds-first-blockchain-coffee-project-cd04fff9e510.

Nash, K. (2007). *Beyond Peter Pan: How ConAgra's pot pie recall bakes in hard lessons for supply chain management*. CIO. Accessed April 12, 2019, from https://www.cio.com/article/2437837/beyond-peter-pan%2D%2Dhow-conagra-s-pot-pie-recall-bakes-in-hard-lessons-for-supply-.html

Opara, L. U. (2003). Traceability in agriculture and food supply chain: A review of basic concepts, technological implications, and future prospects. *Food, Agriculture & Environment, 1*(1), 101–106.

Opara, L. U., & Mazaud, F. (2001). Food traceability from field to plate. *Outlook an Agriculture, 30*(4), 239–247.

OwlTing. (2017). *OwlChain' Launched as the World's first food blockchain provenance system*. Cision PR Newswire. Accessed May 01, 2019, from https://www.prnewswire.com/news-releases/owlchain-launched-as-the-worlds-first-food-blockchain-provenance-system-300464467.html

Petersen, M., Hackuys, N., & See, V. B. (2018). Mapping the sea of opportunities: Blockchain in supply chain and logistics. *Information Technology., 60*, 1–8.

Pizzuti, T., Mirabelli, G., Sanz-Bobi, M. A., & Gomez-Gonzalez, F. (2014). Food track & trace ontology for helping the food traceability control. *Journal of Food Engineering, 120*(1), 17–30.

Quittem, B. (2018). *5 Blockchain projects revolutionizing the supply chain management industry*. Invest in Blockchain. Accessed November 14, 2018, from https://www.investinblockchain.com/supply-chain-blockchain-projects/

Research and Market. (2020). *Blockchain in agriculture and food supply chain market by application (product traceability, payment and settlement, smart contracts, and governance, risk and compliance management), provider, organization size, and region - Global forecast to 2025*. Research and Markets. Accessed February 03, 2021, from https://www.researchandmarkets.com/reports/5203901/blockchain-in-agriculture-and-food-supply-chain

Ringsberg, H. (2014). Perspectives on food traceability: A systematic literature review. *Supply Chain Management, 19*(5/6), 558–576.

Roth, A. V., Tsay, A. A., Pullman, M. E., & Gray, J. V. (2008). Unravelling the food supply chain: Strategic insights from China and the 2007 recalls. *Journal of Supply Chain Management, 44*(1), 22–39.

Saberi, S., Kouhizadeh, M., Sarkis, J., & Shen, L. (2018). Blockchain technology and its relationships to sustainable supply chain management. *International Journal of Production Research, 57*(7), 2117–2135.

Sample, I. (2011). E. coli outbreak: German organic farm officially identified. *The Guardian*. Accessed October 30, 2018, from https://www.theguardian.com/world/2011/jun/10/e-coli-bean-sprouts-blamed

Sawtooth. (2018). *Bring traceability and accountability to the supply chain through the power of Hyperledger Sawtooth distributed ledger technology*. Accessed November 18, 2018, from https://sawtooth.hyperledger.org/examples/seafood.html

ShipChain. (2018). *Disrupting transport & logistics on the blockchain*. ShipChain. Accessed October 30, 2018, from https://shipchain.io

Sims, T. (2018). *Are consumers food safety savvy in 2018?* Food Insight. Accessed October 30, 2018, from https://foodinsight.org/are-consumers-food-safety-savvy-in-2018/

Stanley, A. (2018). Ready to rumble: IBM launches food trust blockchain for commercial use. *Forbes*. Accessed October 19, 2018, from https://www.forbes.com/sites/astanley/2018/10/08/ready-to-rumble-ibm-launches-food-trust-blockchain-for-commercial-use/#43ee334b7439

Steenhuysen, J. (2009). U.S. peanut firm expands salmonella recall. *Reuters*. Accessed October 30, 2018, from https://www.reuters.com/article/us-salmonella-usa/u-s-peanut-firm-expands-salmonella-recall-idUSTRE50Q6RC20090129

Stevens, G. C. (1989). Integrating the supply chain. *International Journal of Physical Distribution & Materials Management, 19*(8), 3–8.

Storer, M., Hyland, P., Ferrer, M., Santa, R., & Griffiths, A. (2014). Strategic supply chain management factors influencing agribusiness innovation utilization. *The international Journal of Logistics, 25*(3), 487–521.

Storoy, J., Thakur, M., & Olsen, P. (2013). The tracefood framework – Principles and guidelines for implementing traceability in food value chains. *Journal of Food Engineering, 115*(1), 41–48.

Su, S. I. I., Gammelgaard, B., & Yang, S. L. (2011). Logistics innovation process revisited: Insights from a hospital case study. *International Journal of Physical Distribution & Logistics Management, 41*(6), 577–600.

Sunny, S. (2018). *Blockchain in Retail: 7 ways Blockchain technology benefits the retail sector*. Smartereum. Accessed October 27, 2018, from https://smartereum.com/39152/blockchain-in-retail-7-ways-blockchain-technology-benefits-the-retail-sector/

SupplyChain247. (2016). *Walmart, IBM and Tsinghua University explore the use of Blockchain in the supply chain*. SupplyChain247. Accessed November 20, 2018, from https://www.supplychain247.com/article/walmart_ibm_and_tsinghua_university_explore_the_use_of_blockchain/bitcoin

Sylim, P., Liu, F., Marcelo, A., & Fontelo, P. (2018). Blockchain technology for detecting falsified and substandard drugs in distribution: Pharmaceutical supply chain intervention. *JMIR Research Protocols, 7*(9), 1–12.

Tang, Q., Li, J., Sun, M., Lv, J., Gai, R., Mei, L., & Xu, L. (2015). Food traceability systems in China: The current status of and future perspectives on food supply chain databases, legal support, and technological research and support for food safety regulation. *Bioscience Trends, 9*(1), 7–15.

Telegraph. (2013). *Horse meat scandal: Timeline*. Accessed October 30, 2018, from https://www.telegraph.co.uk/foodanddrink/9857136/Horse-meat-scandal-timeline.html

Tian, F. (2016). An Agri-food supply chain traceability system for China based on FRID& Blockchain technology. In *2016 13th international conference on service systems and service management (ICSSSM)*, pp. 1–6.

Tian, F. (2017). A supply chain traceability system for food safety based on HACCP, Blockchain & Internet of Things. In *2017 International Conference on Service Systems and Service Management*, pp. 1–6.

Tradelens. (2018). *Digitizing the global supply chain*. Tradelens. Accessed October 30, 2018, from https://www.tradelens.com

Trienekens, J., & Zuurbier, P. (2008). Quality and safety standards in the food industry, developments and challenges. *International Journal of Production Economics, 113*(1), 107–122.

Trienekens, J. H., Wognum, P. M., Beulens, A. J. M., & Vorst, J. G. A. J. V. D. (2012). Transparency in complex dynamic food supply chains. *Advanced Engineering Informatics, 26*(1), 55–65.

Tyco Integrated Security. (2012). Recall: The food industry's biggest threat to profitability. *Food Safety Magazine*. Accessed November 25, 2018, from https://www.foodsafetymagazine.com/signature-series/recall-the-food-industrys-biggest-threat-to-profitability/

Verhoeven, P., Sinn, F., & Herden, T. T. (2018). Examples from blockchain implementations in logistics and supply chain management: Exploring the mindful use of a new technology. *Logistics, 2*(3), 1–19.

Verified Market Research. (2020). *Global blockchain in agriculture and food supply chain market size by geographic scope and forecast*. Verified Market Research. Accessed February 03, 2021, from https://www.verifiedmarketresearch.com/product/blockchain-in-agriculture-and-food-supply-chain-market/

Waldmeir, P. (2008). Chinese toxic milk scandal spreads. *Financial Time*. Accessed November 14, 2018, from https://www.ft.com/content/2a9a21b2-8418-11dd-bf00-000077b07658

Waltonchain. (2018). *About Waltonchain*. Waltonchain. Accessed November 18, 2018, from https://www.waltonchain.org

Wheaton, O. (2018). *Five dead as E. coli outbreak spreads across the US in contaminated lettuce*. Independent. Accessed November 14, 2018, from https://www.independent.co.UK/news/world/americas/e-coli-outbreak-USA-romaine-lettuce-virus-dead-a8380311.html

WHO. (2019). *Drinking-water*. WHO. Accessed April 12, 2019, from https://www.who.int/news-room/fact-sheets/detail/drinking-water

WHO. (2020). *Food safety*. World Health Organization. Accessed May 12, 2020, from https://www.who.int/news-room/fact-sheets/detail/food-safety

Wognum, P. M., Bremmers, H., Trienekens, J. H., & van der Vorst, J. G. A. J. (2011). Systems for sustainability and transparency of food supply chains – Current status and challenges. *Advanced Engineering Information, 25*(1), 65–76.

WTO. (1998). *Understanding the WTO agreement on sanitary and phytosanitary measures*. World Trade Organization. Accessed May 12, 2020, from https://www.wto.org/english/tratop_e/sps_e/spsund_e.htm

WWF. (2018). *New blockchain project has potential to revolutionise seafood industry*. WWF. Accessed October 19, 2018, from https://www.wwf.org.nz/media_centre/?uNewsID=15541

Yakovleva, N. (2007). Measuring the sustainability of the food supply chain: A case study of the UK. *Journal of Environmental Policy and Planning, 9*(1), 75–100.

Yiannas, F. (2018). A new era of food transparency powered by blockchain. *Innovations: Technology, Governance, Globalization, 12*(1–2), 46–56.

Zhao, J. L., Fan, S., & Yan, J. (2016). Overview of business innovations and research opportunities in blockchain and introduction to the special issue. *Financial Innovation, 2*(28), 1–7.

Zwanenburg, J. V. (2018). *What is VeChain (VEN)*. Invest in Blockchain. Accessed November 13, 2018, from https://www.investinblockchain.com/what-is-VeChain/

Chapter 2
Literature Review

This chapter explores the previous literature to gain knowledge relating to the research questions. Therefore, the structure of the literature review chapter is as follows: food supply chain, blockchain, traceability, and theory basis. The *food supply chain is* the first key concept to be addressed as the supply chain has a significant impact on food quality and safety. The characteristics and a few current influencing trends are presented. *Blockchain* follows as the second key concept, as it is the main technological innovation that is investigated and examined in this project. Apart from general information of blockchain, four major revolutionary features are discussed in depth. Next, *traceability* is discussed. This is one of the main areas that can be influenced by blockchain applications. In this section, the motivations and benefits of traceability are explained. The fourth section, the *theories*, mainly focuses on the innovation process model and the practice-based view. These theories are explained in depth and linked up with blockchain applications. The last section presents a content-based analysis of previous research on blockchain in the food supply chain. The analysis generates six benefits and four categories of barriers.

2.1 Food Supply Chain Management

2.1.1 Supply Chain Management

There are various supply chain definitions depending on the industry characteristics. In general, a supply chain describes

> the sequence of processes and activities involved in the complete manufacturing and distribution cycle—this could include everything from product design through materials and component ordering through manufacturing and assembly and onto warehousing and distribution until the finished product is in the possession of the final owner (Franks, 2000, p. 152).

C. Zhang et al., *Blockchain Applications in Food Supply Chain Management*,
Contributions to Management Science,
https://doi.org/10.1007/978-3-031-27054-3_2

The supply chain usually has a large scope that starts with the raw material supply and ends at the point of consumption (Christopher, 2016; Stevens, 1989). It involves more than just physical material flows; it also involves management that includes procurement, supplier management, inventory management, customer service, transports distribution, and other features (Stevens, 1989). In this case, the supply chain is one of the important areas that can be improved to control food risks and increase company performance (Lambert & Cooper, 2000).

A supply chain is "the connected series of activities which is concerned with planning, coordinating and controlling material, parts and finished foods from suppliers to the customer" (Stevens, 1989, p. 3). Mentzer et al. (2001, p. 4) suggested that the supply chain is "a set of three or more entities directly involved in the upstream and downstream, flows of products, services, finance and/or information from a source to a customer". According to Lambert and Cooper (2000, p. 66), "Supply chain management is the integration of key business processes from end user through original suppliers that provides products, services, and information that add value for customers and other stakeholders". These definitions highlight the complexities of the supply chain and the importance of effective information flow. Many researchers also proposed that information flow is one of the important components for supply chain management, as it has a significant influence on the efficiency of the supply chain (Christopher, 2016; Cooper et al., 1997; Grimsdell, 1996; Grunert, 2005; Lambert & Cooper, 2000; Mentzer et al., 2001; Kim et al., 2015; Korpela et al., 2017; Opara, 2003; Spekman et al., 1998; Stevens, 1989).

The supply chain has been transformed from supply-driven to demand-driven (Christopher, 2016; Folkerts & Koehorst, 1997). This means that one of the major supply chain objectives is to satisfy customer demands. Supply chain management aims to satisfy customers by offering high-quality services and economic value (Christopher, 2016; La Londe, 1997; Lambert & Cooper, 2000; Mentzer et al., 2001; Stevens, 1989). In other words, the objectives of the supply chain are "quality, speed, dependability, cost and flexibility" which also means "getting the product in the right condition, in a timely manner and at the lowest possible costs" (Kshetri, 2018, p. 81).

All the activities and management that form a supply chain are interdependent, which means that one process can have an impact on another, due to the interconnected nature of the chain (Folkerts & Koehorst, 1997; Stevens, 1989). This also means that one process collapse can have negative consequences on the whole supply chain; for instance, on performance, efficiency, and effectiveness. The supply chain can be very complex with a broad range of uncertainties, particularly after the trend of globalisation (Christopher, 2016). Managing uncertainties and risk reduction are considered as supply chain objectives (Kshetri, 2018; Lambert & Cooper, 2000). Davis (1993) proposed three uncertainties: suppliers, manufacturing, and customer demand. More specifically, the supply chain uncertainties can include inaccurate demand forecasting and inventory control, damage brand reputation and relationships with stakeholders, and cause unnecessary wastes and costs, as well as unreliable supply sources (Christopher, 2016; Christopher et al., 2011; Cousins et al., 2004).

To achieve supply chain objectives, it is important to pay attention to supply chain efficiency. Christopher (2016, p. 17) presented a competitive advantage formula: *the product excellence multiplied by process excellence equal to competitive advantage*. The core process includes customer management, order fulfilment, and supplier development, to name some. In this case, apart from physical products flow, information sharing, and communication are also part of the efficient supply chain management requirements that were suggested by previous researchers (Grimsdell, 1996; Mentzer et al., 2001). By analysing case studies about how companies managed, Dani and Deep (2010) pointed out the importance of communication and collaboration between supply chain partners. Christopher (2016) and Lambert and Enz (2017) also suggested that supply chain management is about managing relationships with customers and suppliers. There are many innovative information systems that are used in food supply chains and these are introduced in the following sections.

2.1.2 Food Supply Chain

The food industry generally involves the processing, manufacturing, and supply of food. The food supply chain is more precisely defined by Kuo and Chen (2010, p. 559) as

> the total supply process from agricultural production, harvest or slaughter, through primary production and/or manufacturing, to storage and distribution to retail sale or use in catering and by consumers.

In recent years the food supply chain has become more dominated by customer demand and retail interest (Christopher, 2016; Yakovleva & Flynn, 2004). Therefore, a food supply chain is also defined as

> a set of interdependent companies that work closely together to manage the flow of goods and services along the value-added chain of agricultural and food products, in order to realize superior customer value at the lowest possible costs (Folkerts & Koehorst, 1997, p. 11).

Different food products have different characteristics and require specific food supply chains (Dani & Deep, 2010; Roth et al., 2008). In general, the food supply chain is different from other supply chains due to the food characteristic; it is often fragile and very complex (Dani & Deep, 2010; Folkerts & Koehorst, 1997; Trienekens & Zuurbier, 2008; Wognum et al., 2011; Yu & Nagurney, 2013). Complexity of the food chain and the large scale of the food chain can lead to supply chain vulnerability (Kendall et al., 2019). There are a few factors that make food supply chains unique and complex. First, the quality of food products changes continuously within the supply chain. Food products tend to be perishable and contain active materials, which means that food products often come with exact expiration dates (Wognum et al., 2011). The food value tends to be vulnerable, particularly for perishable and fresh products, which require special storage and transport conditions (Trienekens & Zuurbier, 2008). Losing temperature control for temperature-sensitive products during delivery, for example, can significantly shorten the products' life. However, even by the most advanced preserving skills, the quality of food products still tends to decrease over time. Fresh products such as fruits and meats, therefore, draw more attention regarding their freshness and quality. The freshness is also a competitive element for food retailers, which requires better information flow and close collaboration within the supply chains (Ting et al., 2014; Wognum et al., 2011; Yu & Nagurney, 2013).

The products' handling processes including producing, preserving, and delivering require extra attention (Ting et al., 2014). Different food products have different recipes and require various ingredients and components (Wognum et al., 2011). As there is a growing trend of "ready-to-eat" and processed food (Yakovleva, 2007), a variety of mixed ingredients are very normal during the food-producing process (Trienekens et al., 2008). Moreover, food ingredients are from all over the world due to globalisation and outsourcing. In this case, globalisation makes the food supply chain more fragile (Christopher, 2016). Even small mistakes during any steps in the supply chain can have negative consequences including quality failure, or more serious consequences such as food-borne diseases, food poisoning, and contamination. Trienekens and Zuurbier (2008) proposed a diagram for the influencing factors in the fruit supply chain, from planting to retail. For instance, the choices of pesticides in the fruits' growing stage can affect the fruits' safety. Handling methods and storing conditions can impact the fruits' retailing stage. Food quality refers to the storage conditions such as temperature and duration, and also needs to consider the food characteristics. The products' failure not only harms human health, but incurs expenses such as from food recall or sales loss (Dani & Deep, 2010). Christopher (2016, p. 220) proposed a supply chain risk function: *supply chain risk equals to the probability of disruption multiplies the impacts*. Therefore, extra attention is needed in every process of the food supply chain. Responsiveness is one of the key elements of ensuring a successful supply chain performance (Handfield & Bechtel, 2002). High responsiveness and speed information exchange are particularly required in food supply chains to deal with any food hazards and food recalls because speed is crucial to minimise the consequences of risks (Dani & Deep, 2010). Information sharing and communication are the key enablers for supply chain flexibility (Christopher, 2016). Meanwhile, the integrity of the supply chain is again highlighted with certain quality assurance practices.

Second, the food supply chain tends to have many entities involved vertically and horizontally (Pearson et al., 2019; Wognum et al., 2011). In the *vertical* supply chain, the stakeholders are at different stages of the supply chain, including raw material suppliers, producers, distributers, retailers, and customers. *Horizontally*, in each process of the vertical supply chain, more than one company can provide similar products. For example, food producers can receive similar ingredients from different suppliers. The trend of global trade brings more activities, processes, and stakeholders on board. The geographic (location, transports), economic (wealth, income, finance), cultural (language, religious belief, race, social norms) and legislative (policy, history) distance can make the food supply chain even more complex

(Dani & Deep, 2010; Ghemawat, 2001). Ghemawat (2001) suggested that food products such as meat, sugar, cereals, and fruits are all sensitive to different dimensions of distances. For example, meat and cereal can be very sensitive to both different cultures and economies. Sugar can be very sensitive to both political and geographic distances. Perishable food such as fruits are rather sensitive to geographic distances.

Christopher (2016) listed eight sources of supply chain complexity: network complexity (supply chain nodes and links), process complexity (supply chain activities, process, and steps), range complexity (products and service variants), product complexity (products design), customer complexity (service and products customisation), supplier complexity (size of the supplier base), organisational complexity (business structure, management structure), and information complexity (visibility, data volume, forecasting accuracy, data exchange). Among all the complexities, information complexity is contributed to by the other sources of complexity (Christopher, 2016). Due to the complex nature of the food supply chain, collaboration and coordination between stakeholders are significantly important, not only for achieving supply chain efficiency but also for giving quick responses to risks (Christopher, 2016; Dani & Deep, 2010; Trienekens & Zuurbier, 2008).

2.1.3 The Trends of Food Supply Chain Management

Researchers (e.g., Christopher, 2016; Dani & Deep, 2010; Roth et al., 2008) identified a few trends that are affecting the food supply chain: these are globalisation, consolidation, commoditisation, and sustainability. *Globalisation* is recognised by many studies (Christopher, 2016; Dani & Deep, 2010; Folkerts & Koehorst, 1997; Roth et al., 2008; Trienekens & Zuurbier, 2008; Yakovleva, 2007). It is about global sourcing including ingredients, packaging, and food processing (Dani & Deep, 2010). The negative impact of globalisation can be the lack of local resources or the attempt to reduce costs (Folkerts & Koehorst, 1997; Roth et al., 2008). Yakovleva (2007) applied the UK food supply chain as an example and suggested that domestic supply is not able to satisfy domestic demand, which leads to the need to explore outside markets. Either way, it can lead food companies to explore other places for better resources or better prices. Globalisation requires continuous information flow and close collaboration between supply chain partners. According to Lambert and Cooper (2000), the global supply chain should ideally have synchronised customer demand and production rates.

Consolidation refers to the integration by multiple companies to achieve economies of scale (Dani & Deep, 2010; Folkerts & Koehorst, 1997; Lindgreen & Hingley, 2003; Kim et al., 2015; Roth et al., 2008; Yakovleva, 2007). This means that, by merging or integrating with other similar companies into one big company, the company can gain more market share and take control of the supply. The horizontal consolidation has led to market dominance by a few large companies such as Nestle, Coca-Cola, and Cadbury while, for small companies, it imposes entry barriers to the market. Vertical consolidation refers to integration between companies in multiple stages within a supply chain to reduce costs. Trienekens and Zuurbier (2008) pointed out that companies achieve economies of scale and reduce costs by reducing the number of suppliers. Lindgreen and Hingley (2003) also used the giant UK retailer TESCO as an example to discuss the impact of the backward vertical integration, from retailer to supplier integration. By vertical consolidation, companies can have more control over the supply chain and make the supply chain more efficient. Christopher (2016) also suggested that long-term and beneficial relationships with stakeholders can bring advantages to companies. In all kinds of integration, information flow is crucial to improve supply chain performance (Handfield & Bechtel, 2002; Lambert & Cooper, 2000).

Commoditisation refers to the food products with value-added such as organic and fair trade that are becoming commodities. Due to the trend of consolidation, value-added products can be produced on a large scale to satisfy customer demand. For example, all the leading food brands mentioned above (Nestle, Coca-Cola, Cadbury) offer certain value-added food ranges such as organic, low-fat, low-sugar, and fair trade. The commoditisation trend can not only provide the customer with more choices and meet their special needs on products; companies can also gain more profits from the products. According to Roth et al. (2008) and Mai et al. (2010), customers tend to pay more on value-added products. However, the authenticity of products such as organic, calcium-added, low fat and low sugar cannot be easily detected. In this case, customers normally rely on the information that companies choose to provide, which can include product labels and company websites. This can lead to information asymmetry and trust issues between customers and companies. Although many third parties such as governments and international groups provide regular checks and feedback on the products' authenticity, this does not exclude the possibility of undetected dishonest behaviours and bribery. From the consumer perception study by Grunert (2005), quality labels by third parties are not powerful incentives for customers. This means that customers do not believe that the quality labels can present the products' quality and have little confidence in using them as buying cues. Therefore, traceability is necessary as a tool to ensure the food companies act faithfully on value-added products (Costa et al., 2013).

The food supply chain is not only expected to deliver valuable products at the right time to the customers, but also to achieve *sustainability* in the three dimensions of *environment* (pollution, water, climate change), *economy* (finance, profitability, costs), and *society* (health, human rights). Sustainable development is defined as "a development that meets the needs of the present without compromising the ability of future generations to meet their own needs" (WCED, 1987). Seuring and Muller (2008, p. 1700) conducted a literature review and defined sustainable supply chain management as.

> the management of material, information and capital flows as well as cooperation among companies along the supply chain while taking goals from all three dimensions of sustainable development, economic, environmental and social, into account which are derived from customer and stakeholder requirements.

Of the three dimensions, the environmental dimension has attracted the most research interest, as Seuring and Muller (2008) suggested. Otherwise, businesses tend to focus more on economic sustainability and strive to achieve constant profits and financial growth. In recent years, however, the social dimension is gaining more attention. People care more about food nutrition, food quality, food-producing environment, human rights, and animal welfare, among other issues. Food quality and safety is one of the concerns in the social dimension. This is due to the increasing number of food hazards, food poison news, and food recalls. To achieve sustainability, third parties are necessary for companies to develop sustainable supply chain activities, as the regulations and verifications can encourage companies to turn the sustainability concept to actual actions. This also means that sustainability requires all stakeholders to become involved and collaborate to achieve a sustainable supply chain (Seuring & Muller, 2008).

2.2 Blockchain

2.2.1 Blockchain Description

Once food disease happens, product recalls should also be made quickly to avoid further damage and loss. This requires companies to have good traceability systems that can flexibly handle all product recall crises. However, as the food supply chain is complex, involving multiple stakeholders, the ability to trace products all the way back to their origins can pose a huge workload (Roth et al., 2008). To adjust the information inequality between stakeholders and trust issues within the supply chain, blockchain is one of the promising technologies that is helpful and is expected to make revolutionary changes in supply chain management (Kouhizadeh & Sarkis, 2018; Queiroz et al., 2019). It is predicted to significantly improve information transparency and traceability systems (Petersen et al., 2018; Queiroz et al., 2019). It is also a solver for trust issues and brings public confidence to the food industry. Information systems were found to be helpful in enhancing company information flow and supply chain performance (Kim et al., 2015) (Table 2.1).

2.2.2 Blockchain Features

To summarise, blockchain has the characteristics of decentralisation, smart contract, immutability, and security.

2.2.2.1 Decentralisation

One of the main customer concerns is food safety issues and the negative impacts on health (Yakovleva, 2007). As many food incidents are caused by lack of visibility in the supply chain and misuse of power by the central authorities, such as the Sanlu

Table 2.1 Blockchain descriptions

Researchers	Description
Seebacher and Schuritz (2017, p. 14)	"A distributed database, which is shared among and agreed upon a peer-to-peer network. It consists of a linked sequence of blocks, holding timestamped transactions that are secured by public-key cryptography and verified by the network community. Once an element is appended to the blockchain, it cannot be altered, turning a blockchain into an immutable record of past activity".
Gupta (2017, p. 5)	"A shared, distributed ledger that facilitates that process of recording transactions and tracking assets in a business network. An asset can be tangible—a house, a car, cash, land—or intangible like intellectual property, such as patents, copyright, or branding".
Diamond Time-lapse (2018)	"Blockchain, also known as distributed ledger technology, is a decentralised distributed database that maintains a continuously-growing list of records called 'blocks', that are secured from tampering and revision without the need for central authority".
Weber et al. (2017, p. 65)	"A blockchain system can be thought of as an append-only, public ledger that keeps track of transactions made by participants".
Tian (2016, p. 3)	"the essence of the blockchain is a technical scheme of reliable dataset which is collectively maintained by the way of decentralised and trustless method".
Lin et al. (2017, p. 60)	"Blockchain is essentially a decentralised distributed ledger database which consists of a chain formed by cryptographic data blocks, each data block contains transaction data that already been confirmed".
Petersen et al. (2018, p. 1)	"Blockchain is an emergent technology concept that enables the decentralised and immutable storage of verified transaction data".
Kouhizadeh et al. (2021, p. 3)	"Blockchain is defined as decentralized ledgers that contain transactions as data blocks; with blocks linked to their predecessors by a cryptographic pointer".
Stranieri et al. (2021, p. 1)	"Blockchain technology (BCT) is a digital technology which allows transaction and information flows without the need of intermediaries".

milk scandal in China, this thus emphasised the importance of transparency and trust within a supply chain. Traditionally, the food supply chain is centralised, which means that the supply chain relies on central authorities to monitor and control flows. For information flow, centralisation can potentially encourage companies to release a certain amount of information that benefits them. Central authorities can also become the targets of bribery and help companies to cover up dishonest behaviours. From the consumers' point of view, food producers tend to provide partial information that benefits themselves, which can potentially mislead the public (Mao et al., 2018). Central authorities, that should stand for consumers' rights, can also be tempted by bribes. From companies' points of view, it is hard to be familiar with all the suppliers and trust all supply chain partners (Christopher, 2016; Wognum et al., 2011). It is not uncommon that one company has hundreds of lower-tier suppliers providing materials to its upstream supply chain, while all the suppliers are independent but also have interdependent relationships (Christopher, 2016).

Under normal conditions, companies use certain ways such as paper contracts to regulate their partner's behaviours and maintain trust. However, the willingness to cooperate with each other can highly depend on the distribution of power along the supply chain (Handfield & Bechtel, 2002). The level of collaboration is usually based on trust and personal relationships. In this case, to build long-term relationship with suppliers, trust is one of the key elements to form a good stakeholder relationship. By building a theoretical framework, Handfield and Bechtel (2002) suggested that trust has a significant impact on responsiveness and supply chain performance.

The centralised supply chain can be very monopolistic and lead to information asymmetry. Stakeholders have unequal power in a food supply chain. Thus, the absolute power of a single stakeholder over another can lead to information fraud, information manipulation, and corruption. All these can further cause trust issues between stakeholders (Tian, 2016, 2017). The Sanlu milk scandal is one of the typical examples of where an opaque centralised supply chain and irresponsible authorities led to food quality failure. Therefore, transparency and trust issues within a supply chain need to be solved. Not only are consumers demanding more information about the products they purchase, but food companies also need to have more knowledge of their suppliers (Wognum et al., 2011).

To avoid the misuse of power by authorities, blockchain technology can provide stakeholders with trust and transparency of the information by decentralisation, and prevent the potential vulnerability from centralisation (Kennedy et al., 2017). The blockchain-based solution allows everyone related to the supply chain to access and manage information including temperature control or sources in real time (IBM, 2018). Customers will be able to check food origins, get to know the stories behind the purchasing, the involved human labour ethicality, and transportation information, as well as other points. Grunert (2005) proposed that products' origins can be considered as a quality cue by customers. This suggests that information about products such as origins can strongly affect customers' buying decisions. The open-to-the-public and simultaneous information flow not only provides transparency of the supply chain and reduces the central authority power, but also reduces risks of product fraud and low quality. Blockchain can also be used as a marketing tool as the transparency of products can improve public confidence, enhance customer loyalty, and attract new customers (Galvez et al., 2018).

2.2.2.2 Security

The decentralisation feature can bring network security. As all transactions are validated by user to solve mathematical algorithm mechanism, it can potentially secure the data by the same method (Tian, 2016) and can significantly reduce incidents of cyber-attack (Zhao et al., 2016). As blockchain is a ledger without a central power, cyber-attacks only happen when half or more (51%) of the power is held by the attacker. If one single node/unit controls half or more (51%) of processing power, it can dominate all the other nodes (Zhao et al., 2016). Therefore, the more nodes/units there are in blockchain, the more stable it is. Blockchain can also protect privacy by separating ledgers as public (open), federated or consortium, or private (permissioned) while ensuring information sharing (Gausdal et al., 2018; Lin et al., 2017; Tseng et al., 2018; Wu et al., 2017). The differences between private and public ledgers are the requirements on data access and validation methods (Gausdal et al., 2018). The public ledger requires less permissions, while the federated/consortium ledger needs more confirmation on every transaction. The private ledger is similar to the federated/consortium ledger in that it is limited to one company's operations.

2.2.2.3 Smart Contract

Another major benefit brought by blockchain is the use of the smart contract. Smart contracts are computer codes that are based on the agreement of all relevant verified companies/users. As it is a programmed contractual agreement between parties, it digitalises all the paper documents, runs transactions, and keeps information in a more secure way (IBM, 2018; Seebacher & Schuritz, 2017). "It's a contract that self-executes, and the contract handles the enforcement, the management, performance, and the payment", according to Don Tapscott (Blockgeeks, 2018). The transactions run automatically by blockchain when certain conditions are met. For example, the smart contract can automatically send payment to the client once products are received. In this case, by using the smart contract, it not only reduces the number of third parties involved but also increases the transaction efficiency. Once smart contracts are agreed and settled, no single entity, central power, or third parties can change it. This also means that there is no need for trust between parties using the smart contract as it is run by programmed code (Jeppsson & Olsson, 2017). Unlike paper documents, smart contacts can save more time and labour on document-processing as it can be triggered and run automatically in certain conditions (Zhao et al., 2016). It can also reduce human errors, and make transactions safer and quicker (Verhoeven et al., 2018). For example, in 2014, Maersk found that over 30 people and organisations were involved when shipping a container of roses and avocado from Kenya to the Netherlands (Park, 2018). It also took 34 days including 10 document-processing days to complete a whole shipping activity, which does not include missing documentation-caused delay and time extension (Park, 2018). In this case, blockchain can achieve maximum digitalisation, save processing time, increase efficiency, and reduce unnecessary costs (Kshetri, 2018).

In addition, tokenisation is a function of blockchain that usually combines with the smart contract. Token is a digital form of asset that is generated on blockchain. In other words, token is an algorithm that represents certain assets. Token can be used to secure certain assets because of its uniqueness.

2.2.2.4 Immutability

In the blockchain ledger, all the information will be verified and stored securely during and after transactions, and data can also be checked by relevant users at any time. Once a transaction has been done, it can be accessed and retrieved by verified users. However, it cannot be changed. The real-time information flow and information immutability also allow companies to trace back and check the producing processes whenever needed, which is particularly necessary in cases of product recall. Once food incidents happen, product recall can be done quickly to avoid further damage to the organisation. However, before blockchain is widely adopted in the food industry, product recall can be very difficult, can take a long time and can be costly. Food recalling requires multiple procedures such as sample collection and risks evaluation, and the processes can take over 100 days in some cases (Chamlee, 2016). According to the Food Marketing Institute, in 2012, the average direct cost for a food recall was about $10 million in the USA, as well as the indirect costs such as lawsuits, or loss of reputation (Tyco Integrated Security, 2012). Therefore, in 2017, Walmart founded a food safety and collaboration centre in China and worked with IBM and Tsinghua University to improve food safety and address the food crisis by improving the traceability system (Loria, 2017). The pilot study investigated pork in China, and mangos in the US and found that products' tracing time can be significantly reduced by using blockchain, from days/weeks to seconds (Aitken, 2017; IBM, 2017). Therefore, immutability is significantly useful to improve the traceability system to find products' ownership transfers within the supply chain (Verhoeven et al., 2018). Apart from time efficiency in product recall, the immutable database can also safeguard products' authenticity and prevent fraudulent activity (Petersen et al., 2018; Verhoeven et al., 2018). From the raw material suppliers, transportation to retailers, every process within the supply chain can be traced back in blockchain. Frank Yiannas, Walmart's Vice president for food safety, asserted:

> blockchain technology enables a new era of end-to-end transparency in the global food system—Equivalent to shining a light on food ecosystem participants that will further promote responsible actions and behaviours. It also allows all participants to share information rapidly and with confidence across a strong trusted network. This is critical to ensuring that the global food system remains safe for all. (Cited in Agres, 2017).

Many researchers examined blockchain applications and the impact on the food supply chain. Most reported positive attitudes of blockchain utilisation and acknowledged its importance to the food industry. More specific analysis can be found in the following content-analysis section.

2.3 Traceability

2.3.1 Traceability Definitions

Due to the globalisation of modern businesses and the economy, food supply chain integrity has been extended from safety concerns to include origin fraud and quality concerns. The international food trade is often disrupted by disputes over food safety and quality requirements (FAO, 2003) and food-borne diseases are increasing on a global scale. The World Health Organization (WHO) estimated that more than 200 diseases caused by unsafe food products can lead to 600 million people falling ill each year (WHO, 2020). Customers are increasingly demanding more information on the food products they purchase (Wognum et al., 2011). In addition, food-borne disease can damage the economy as it is counter-productive and imposes a substantial burden on the healthcare systems. To ease the global concerns for food safety and to protect the health of consumers and the economy, it is necessary to build an effective food control system. Therefore, the traceability system was introduced, and it plays a more important role in the food supply chain as various food insecurity scandals happen more often nowadays.

The earliest definition of traceability can be found in the International Organization for Standardization (ISO, 1994): "the ability to trace the history, application or location of an entity by means of recorded identification". However, this definition is not strictly used for the food supply chain. A few studies have also provided different definitions of traceability that can be found in Table 2.2. In this literature review, Bosona and Gebresenbet (2013, p. 35) provided a more complete definition of the food tractability system:

> food traceability is part of logistics management that capture, store, and transmit adequate information about a food, feed, food-producing is correct animal or substance at all stages in the food supply chain so that the product can be checked for safety and quality control, traced upward, and tracked downward at any time required.

Costa et al. (2013, p. 353) suggested that traceability in the agri-food industry should refer to "the ability to locate an animal, commodity, food product or ingredient and follow its history in the supply chain forward (from source to consumer) or backward (from consumer to source)".

Traceability concerns the vertical supply chain, from the products' origins to the end customers (Wognum et al., 2011). Bosona and Gebresenbet (2013) also asserted that an efficient food traceability system should have three main components:

Table 2.2 The definitions of traceability

References	Definition
ISO 8402 (1994)	"the ability to trace the history, application or location of an entity by means of recorded identification".
Wilson and Clarke (1998, p. 128)	"the information necessary to describe the production history of a food crop, and any subsequent transformations or processes that the crop might be subject to on its journey from the grower to the consumer's plate".
European Union (European Commission, 2002)	"the ability to trace and follow a food, feed, food-producing animal or substance intended to be, or expected to be incorporated into a food or feed. Through all stages of production, processing and distribution".
ISO 9000 (2005)	"the ability to trace the history, application or location of what which is under consideration".
Codex Alimentarius Commission (2015, p. 26)	"the ability to follow the movement of a food through specified stage(s) of production, processing and distribution".

tracing, tracking, and product history information. *Tracing* (backward traceability) is the ability to follow products along the supply chain backward to the origins, while *tracking* (forward traceability) considers the forward follow-up of the products to the end customers. *Products history* means the sufficient and complete movements of products in the supply chain. Ene (2013) also suggested that traceability is about a records system and stakeholder communication. By having a complete and efficient traceability system, it is possible for companies to check the products in any process of the supply chain.

2.3.2 The Motivations for Traceability

A few studies analysed the motivations for food companies to install traceability systems. For example, Golan et al. (2004) used the USA food industry as an example and proposed three main objectives and benefits of the food traceability system: improve supply chain management, achieve food quality control, and increase public confidence and trust. Bosona and Gebresenbet (2013) reviewed 74 studies and summarised five drivers for a traceability system: food safety and quality, regulations and laws, social concerns, economic concerns, and technological concern. Some other papers also proposed drivers for companies implementing traceability system (Canavari et al., 2010; Ene, 2013; Mai et al., 2010; Meuwissen et al., 2003; United Nations Global Compact, 2014; WEF, 2019). Therefore, to combine all these, six main motivations are summarised and listed in the following context: to enhance and address food safety issues; to increase customer trust and buying confidence; to follow the traceability laws; to improve supply chain management; to achieve sustainability, and to encourage technology innovations.

First, to enhance food products' safety and quality, and eliminate food recall scales. Ene (2013) argued that improving food security is the main purpose of traceability. The traceability system can improve visibility in the supply chain as it holds records and information. It can also be used as a strategic tool to encourage suppliers to provide quality products and services (Galvão et al., 2010). Apart from potentially improving products' quality, traceability is also a solution for products' failure which leads to products' recall (Kelepouris et al., 2007) as traceability is an efficient tool for food companies to reduce products failure in the first place. In other words, the traceability system can be used as a proactive strategy. Once product failure happens, adequate information of products can provide great help for companies on tracing and isolating faulty products at fast speed. It, therefore, can reduce further damage to the public through faulty products, and eliminates the products' recall costs. The more sufficient information the traceability system holds, the quicker products' recall can be done and products' failure can be addressed (Golan et al., 2004).

Second, to enhance customer buying confidence and attract more customers to gain economic benefits. Companies can also use traceability as a marketing tool to differentiate themselves from the others in the area of quality assurance (Mai et al.,

2010). Good brand image is a strong cue to attract purchasers as the products' quality can be highly predictive (Grunert, 2005). To balance the revenue and traceability costs, companies can also increase products' prices, as Sun et al. (2017) proposed that customers tend to pay higher prices for traceable food products as they have more consciousness on traceability.

Third, to follow traceability laws and policies. Food products' standardisation or policies are important to protect consumers' rights and eliminate food risks (Aruoma, 2006; European Commission, 2019). As mentioned in Chap. 1, many legislations and policies were published to encourage food companies to implement traceability systems, such as the EU food law, and international organisations for standardisation, etc. Following the laws and regulations, companies are forced to have a certain standard of traceability system for their food products (Wognum et al., 2011). The verifications from third parties can provide customers with a certain level of confidence and trust (Bosona & Gebresenbet, 2013). In some cases, food companies can also benefit financially by installing traceability from third parties (Bosona & Gebresenbet, 2013).

Fourth, to improve supply chain management and demand management. One of the main issues of the supply chain is uncertainty, which is about losing control over supply and demand. The "bullwhip" effect, for example, is usually caused by an inaccurate forecast on customer demand. As mentioned above, controlling uncertainties and minimising risks are important for achieving supply chain efficiency, and will need good demand management (Lambert & Cooper, 2000). This means that the traceability system is part of making up a good supply chain, which not only helps supply chain managers make decisions on managing resources such as produce, warehouse, and transport, but also enhances the engagements and coordination between supply chain partners (Bosona & Gebresenbet, 2013; Golan et al., 2004).

Fifth, to reduce unnecessary costs and achieve sustainability. As margins are relatively thin in the food industry (Roth et al., 2008), the ability to accurately forecast the demand and provide the right amount of supply can reduce unnecessary costs within the supply chain, which can make a significant difference in competition. Unnecessary costs and wastes between products' movements can be eliminated. Christopher (2016) pointed out that the more complex the supply chain, the less effective the forecasting tools will be. Traceability, therefore, can be used as a competitive strategy (Canavari et al., 2010; Dani & Deep, 2010). On environmental and social sustainability, traceability can help with product loss by using effective packaging and eliminating unethical sources of produce (Bosona & Gebresenbet, 2013; WEF, 2019).

Sixth, apart from the three objectives mentioned above, *advanced technology can also be an important driver for companies to have efficient traceability system* (Bosona & Gebresenbet, 2013). Installing a traceability system can require economic and human labour investments with complicated devices. Although a traceability system is useful and necessary, it also is a huge pressure for the food industry as it has thin profit margins and a large number of small to medium enterprises (SMEs). Therefore, advanced technologies with lower investments and higher effectiveness can revolutionise the traceability system in the food industry and attract more attention. The following sections introduce one of the most promising technologies—*blockchain*.

2.3.3 The Benefits of Traceability

Along with the motivations, the traceability system can also benefit all stakeholders in a supply chain. For companies, the benefits are obvious. Mai et al. (2010) used the fish industry as an example to investigate the benefits of traceability. The benefits are summarised as qualitative and quantitative and include market benefits, recall costs saving, liability costs saving, process improvement saving, and labour cost saving. More specifically, traceability can be used as a marketing tool to enhance customer trust and generate more sales (Bosona & Gebresenbet, 2013; Golan et al., 2004; Mai et al., 2010). As a record-keeping system, traceability can also increase trust between partners, reduce information asymmetry, and encourage suppliers to provide quality materials. The increased visibility of the supply chain can also help companies to manage production and stocks, prevent unnecessary wastes in products' movements, and achieve sustainability (Bosona & Gebresenbet, 2013). As an information-based system, traceability is a cost-efficient and speedy solution for addressing food products' failure by identifying and specifying certain products (Bosona & Gebresenbet, 2013).

Products' recall can lead to multiple negative consequences: customer confidence loss, brand image damage, recalling costs, and product waste. Grunert (2005, p. 381) suggested that food safety is a "sleeping giant", which refers to its strong impact on consumer decision-making during food hazards. While the food industry situation is normal and secure, food safety is usually not the consideration that comes to consumers' minds. There are two main causes for food recall (Dabbene et al., 2014). The first is *manufacturing practice failure* such as wrong labelling, packaging information, and chemical contamination. For example, Italian brand Wish-Bone salad dressing had to recall the mislabelled products that did not state the possible allergens present in the product (Shaw, 2018). Melissa's Spicy Edamame recalled their snack due to the packaging failure (Shaw, 2018). The second main cause is *material failure*, including raw material contamination and allergens. For instance, an Arizona-based meat producer JBS Tolleson recalled a total of 12 million pounds of raw beef due to an outbreak of salmonella (Shaw, 2018). Utz Quality Food issued a chips recall due to undeclared milk contamination (Shaw, 2018). All the examples are from 2018, however; in the same year, nearly 200 other food recalls happened in the USA. The food crisis scandals tend to increase every year.

For customers, traceability is the added value for food products that relates to more precise information (Dabbene et al., 2014). Traceable products can improve supply chain visibility, enhance products quality and authenticity, and help them to make decisions before purchasing (Bosona & Gebresenbet, 2013; Opara, 2003). It can speed up the food recall process and identify the failure products quicker, and launching traceability can further enhance customers' trust in products' safety.

Traceability can also potentially be the incentives of technology development and encourage deep investigations in technological innovations in information system area (Bosona & Gebresenbet, 2013). Cost-efficient traceability systems with lower installing fee and easy using features are also increasing in demand, particularly for SMEs. However, it is worth noting that it is hard to achieve total traceability in food supply chain as the food supply chain can be very complex, particularly for processed food and composed food, as it is very difficult to trace back every single ingredient to its origins (Wognum et al., 2011).

2.3.4 Transparency

Transparency is strongly related to traceability. It is defined as "the degree of shared understanding of and access to product-related information as requested by a supply chain's stakeholders without loss, noise, delay, or distortion" (Wognum et al., 2011, p. 65). Traceability relies on the degree of transparency, which means that the level of transparency can impact traceability efficiency. High level of transparency is required to build efficient traceability (SGS, 2018). Handfield and Bechtel (2002) also proposed that the level of trust in a supply chain can impact the speed of responsiveness and supply chain performance. While traceability focuses on the individual product operational information such as movements and batch-lot data to achieve recall efficiency, transparency concerns more of the products' information such as ingredients, certifications, suppliers, expiration dates, and origins (SGS, 2018). In this case, by having a high level of transparency, it is possible for companies to know more about stakeholders and map the whole supply chain network.

2.3.5 Technology Innovations on Food Traceability

As mentioned above, due to the geographic, cultural, economic, and political distances caused by globalisation, food recall in the global food trade can be more challenging with lower responsiveness (Christopher, 2016; Ghemawat, 2001). For example, Roth et al. (2008) mentioned that after Sanlu contaminated milk caused pets' deaths in the USA, the FDA was not allowed to enter China for weeks to undertake their investigations. The slow reaction to food recall can be largely addressed by implementing an efficient traceability system. To tackle the distance issues, a global traceability system should be developed to be more flexible to suit global food trade and markets (Pizzuti et al., 2014). An efficient traceability system allows people to trace and track the food along the whole supply chain, which can mean a "tremendous technical undertaking requiring sophisticated markers and software" (Roth et al., 2008, p. 32). Many technology innovations had been investigated to improve food quality and safety issues, as well as food recall.

Christopher (2016) suggested that information flow is affected by its collecting conditions and ability to create value. The current traceability system still largely relies on either a paper-based system or isolated internal computer systems (Aung & Chang, 2014; Danovich, 2016; Manos & Manikas, 2010). Paper recording can be time- and labour-consuming. Paper files can also get lost during information exchange in food supply chain. Recording mistakes can be another weakness that often happens due to manual entry errors. In some scenarios, even paper files can rarely be used due to cash-based transactions that largely happen in lower-tier suppliers in developing countries. Some large companies may have an internal traceability system that can handle and store more information. The computer system is surely more efficient than paper recording; however, it may be impossible for the other stakeholders to adopt. This can disrupt stakeholder integrations. In the Walmart blockchain pilot case, the traditional method of tracing products origins is by calling the suppliers layer by layer which can take up to 7 days.

The barcode is an early-stage technology that is widely used on many products. The barcode stores readable data that can be scanned by reader or smart phones. It can hugely reduce manual input errors and save labour costs. The cheap and easy operation of the barcode made it very popular. However, the barcode also has certain issues such as very limited data storage size, and damages easily, etc. (Adaptaliftgroup, 2012). For example, the barcode reading rate is found to be decreased to less than 90% in some situations, and damaged or distorted barcodes can be hard to read (Costa et al., 2013).

The Internet of Things (IoT) is called the fourth industrial generation which has attracted significant interest (WEF, 2019). It is "a global network infrastructure" that can connect different devices and objects by internet, where the objects interact with each other and exchange information (Atlam et al., 2018; Badia-Melis et al., 2015; Xu et al., 2018, p. 2944). Two major innovations of IoT are radio frequency identification (RFID) and wireless sensor networks (WSN). RFID uses tags and readers to identify and store information and achieves efficient products tracking and monitoring by managing information on certain RFID software (Costa et al., 2013; Xu et al., 2018). It is recognised as the "next generation of barcode" as it is more efficient and more reliable than barcodes (Kelepouris et al., 2007, p. 185). It allows a higher information reading rate which can achieve 99.5–100% efficiency, and many RFID tags can be read at the same time. (Adaptaliftgroup, 2012; Badia-Melis et al., 2015; Costa et al., 2013; Kelepouris et al., 2007). The wide use of RFID can largely improve supply chain visibility, traceability, information accuracy, and efficiency. Therefore, many researchers make positive comments on RFID (Christopher, 2016; Costa et al., 2013; Mai et al., 2010; Kelepouris et al., 2007; McMeekin et al., 2006; Wognum et al., 2011).

WSN is a network using sensors to monitor and collect information from the environment, such as humidity, temperature, and vibration, among other aspects (Nagpurkar & Jaiswal, 2015). The ability to monitor environmental conditions on a near-real-time basis is particularly helpful for temperature-sensitive products such as frozen food and fresh food. The combination of the two technologies can achieve consistent condition monitoring (Costa et al., 2013). Combining IoT with blockchain applications is another major trend, which has been investigated in many studies. For example, Kshetri (2018) suggested that blockchain can address IoT challenges such as IoT security. Atlam et al. (2018) also proposed the benefits of integrating blockchain and IoT, such as security, resiliency, privacy, and reliability. Many studies also more specifically investigated the benefits of combining the two technologies in food supply chain management (Leong et al., 2018; Lin et al., 2017; Tian, 2016).

There are many other emerging technologies in monitoring food products in the supply chain. Near Field Communication (NFC) is a more updated version of RFID; it is smaller and allows for short-distance information sharing (Badia-Melis et al., 2015). It has been successfully installed in making instant payments such as mobile wallet. DNA-based application, for instance DNA-barcoding, is a very accurate way to check and store products' information, particularly for meat and seafood products (Badia-Melis et al., 2015; Galimberti et al., 2013). The major problem is cost inefficiency, which may not be suitable for the intensive food industry (Wognum et al., 2011). Shanahan et al. (2009) also suggested the integration of the EPCglobal Network (EPCglobal, 2009) for data sharing. The World Economic Forum (WEF, 2019) proposed the food-sensing technology, which can identify products' structure information (moisture, protein, fat content, etc) and upload this to be analysed for food quality and safety purposes.

Other examples in achieving traceability include, but are not limited to, the use of web-based systems for tracking and tracing agricultural batch products along the food chain (Ruiz-Garcia et al., 2010), the development of a Flexible Tag Datalogger (FTD) to collect environmental data in the wine logistic chain (Mattoli et al., 2009), and using GPS to locate vehicles and plot the optimal delivery route (Tian, 2016). The appropriate technology implementations can improve the data exchanging efficiency, enhance trust, and benefit supply chain stakeholders (Aung & Chang, 2014). However, none of the technologies mentioned above is perfect; challenges also exist alongside the opportunities (Table 2.3).

2.4 A Content-Based Analysis of Blockchain and the Food Supply Chain

2.4.1 Literature Review Method

To answer the research questions, this research adopts a literature review method. There are several review papers on blockchain and supply chain management; however, none has a specific focus on the food supply chain (Wang et al., 2019a, b). These papers have provided the foundation of this research.

Fink (2005, p. 3) defined the literature review as "a systematic, explicit, and reproducible design for identifying, evaluating, and interpreting the existing body of recorded documents". This research adopts a content-based literature review on the

Table 2.3 Traceability methods

Methods	Merits	Drawbacks
Paper files	1.Easy to use	1.Easy to make mistakes or get lost 2. Costs and time-consuming
Internal company system	1. Easier to use than paper recording 2. Lower costs and less time-consuming than paper recording	1. Incompatible for other stakeholders 2. Still need manual input
Barcode	1. Low costs 2. Easy to use 3. Fast speed 4. Can lower the possibility of human errors	1. Can only contain very simple and basic information 2. Only allows line-of-sight scan 3. Distortion or damage on the barcode can destroy the scanning effect
RFID	1. It contains more information than barcode 2. Allows us to read multiple objectives at one time 3. It does not have line-of-sight scan limitations 3. Can re-use the tags 4. The scanning can be done at fast speed automatically	1. It is still more expensive than barcodes with high initial costs 2. Tag collision and reader collision may happen 3. Data manipulation
WSN	1. Can automatically capture information and save labour 2. Flexible network and adaptable to changes.	1. Security issues 2. Nodes need to be charged regularly 3. Low communication speed 4. Sensor network can be distracted
NFC	1. Convenience 2. Allows short-distance real-time information sharing 3. High level of encryption can bring more security in the payment than physical card payments	1. Expensive to purchase and maintain the equipment for small companies 2. Unauthorised access and mobile hacking can still happen
EPCglobal Network	1. Can combine with other IoT such as RFID and barcode. 2. It can manage the dynamic information of specific products	1. Privacy and security issues 2. Data confidentiality
DNA-related technology	1. Information accuracy and reliable 2. It is the only method that allows correct identification of species in complex foods 3. The identification scope is unlimited 4. Highly customisable 5. It has an untargeted nature that allows full knowledge of the DNA content of a food sample	1. High costs 2. It might be different to differentiate the closely related species particularly when the products have been highly processed and lead to DNA fragments.
Blockchain	1. Network security 2. Information sharing leads to transparency 3. Immutable history information 4. Smart contract	1. Immature technology with certain issues including technology scalability, lack of understanding and knowledge, lack of regulations. 2. Requires all related stakeholders to get involved.

application of blockchain in food supply chain management. This research not only thoroughly reviews blockchain, but also considers practical applications in food supply chain management. It aims to provide a more precise and integrated understanding of blockchain and its influences in food supply chain management.

This literature review applies the six-stage refinement process suggested by Durach et al. (2017): (i) define research question, (ii) set inclusion and exclusion criteria, (iii) determine searching databases, (iv) apply criteria, (v) synthesise relevant literatures, and (vi) report findings. Research questions have been provided in the introduction. Therefore, the research keywords are *blockchain* and the *food supply chain*. Web of Science, Scopus and Ebsco are the three online databases that the researcher accessed to search relevant academic literature, as the three databases have a wide range of resources and have been used extensively in supply chain management research. Peer-reviewed journal articles are seen as high-quality communication between research fellows. However, in this research, due to the early stage of blockchain and the limited published articles, other resources such as conference papers, consulting reports, and third-party reports that can provide more updated information were also considered. The initial search generated 137 results from the three databases.

Once duplicated papers were removed and inclusion and exclusion criteria were applied (Table 2.4), 58 papers were left in the final process (Fig. 2.1). The final 58 papers were categorised and evaluated by content analysis method, which is a systematic and objective research method that has been used to quantify phenomena, documents, or communications (Seuring & Gold, 2012).

Descriptive analysis is the first insight into all the papers and is supposed to provide basic information on the content of the selected papers. Among the selected papers, due to the young age of technology, the earliest paper (one out of 58) was released in 2016; nine papers in 2017; 16 papers in 2018; eight papers in 2019; and 24 papers in 2020 and onwards. The time trend shows that blockchain is gaining more attention and interest in the supply chain area. Seventeen out of 58 papers are technology- and innovation-related conference papers (Caro et al., 2018; Lin et al., 2017, 2018; Tian, 2016, 2017; Tse et al., 2017). The papers focus on different aspects of the food supply chain, but 12 out of 56 of the papers focus mainly on *traceability* (Casino et al., 2020; Feng et al., 2020; Pearson et al., 2019; Salah et al., 2019; Sander et al., 2018; Tian, 2016). A few papers also discussed *blockchain*

Table 2.4 Inclusion/exclusion criteria

Inclusion	Exclusion
Published in English language	Published in other languages
Papers focus on food supply chain only	Papers focus on any industry rather than food supply chain industry
Published since 2008 to June 2021	Published before 2008
Papers focus on blockchain	Papers focus on other technologies
Peer review/ conference papers	Business news
Management focus	Technique focus

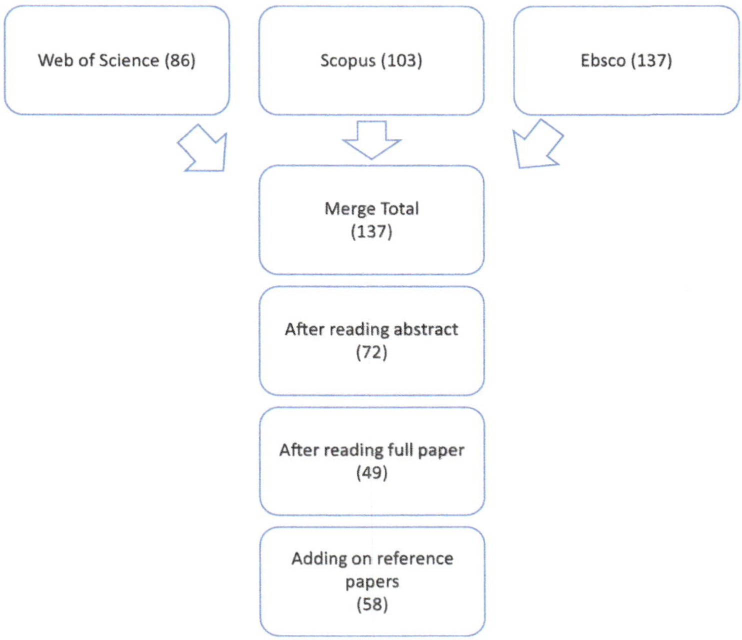

Fig. 2.1 Paper selection process

adoption (Kayikci et al., 2020; Kouhizadeh et al., 2021; Tsolakis et al., 2020; van Hoek, 2019), *supply chain sustainability* (Skender & Zaninovic, 2020; Song et al., 2020) and *supply chain visibility* (Novianti et al., 2020; Rogerson & Parry, 2020). Although all papers focus on food supply chains, the food categories are slightly different. Most papers focused on the agri-food supply chain in general (Caro et al., 2018; CBH Group and AgriDigital, 2017; Faye, 2017; Ge et al., 2017; Hua et al., 2018; Lin et al., 2018; Tian, 2016, 2017; Tripoli & Schmidhuber, 2018; Zhao et al., 2019). A few papers used more specific food supply chains such as the Halal supply chain by Tieman and Darun (2017) and Novianti et al. (2020), the tilapia supply chain by Rejeb (2018), the rice supply chain by Kumar and Iyengar (2017), the fish supply chain by Tsolakis et al. (2020) and Howson (2020), the soybean supply chain by Salah et al. (2019), the dairy supply chain by Casino et al. (2020), and the coffee supply chain by Thiruchelvam et al. (2018). The reviewed papers mainly introduce blockchain and demonstrate its potential by using a conceptual framework (28 out of 58), pilot cases (15 out of 58), three theory paper s (four out of 58), two surveys (two out of 58), eight systematic literature review analysis (eight out of 58), and one report.

All the papers were read thoroughly and highlighted by the different categories of benefits and challenges (Table 2.5). The benefits can be divided into four categories: (i) transparency, (ii) information authenticity, (iii) efficiency, and (iv) sustainability; the challenges are: (i) lack of understanding, (ii) immature technology, (iii)

Table 2.5 List of papers by content analysis

Categories	Papers	Tian (2016)	CBH Group and AgriDigital (2017)	Faye (2017)	Ge *et al.* (2017)	Hackius and Petersen (2017)	Kumar and Lyengar (2017)	Lin *et al.* (2017)	Tieman and Darun (2017)	Tian (2017)	Tse *et al.* (2017)	Bermeo-Almeida *et al.*(2018)	Caro *et al.* (2018)	Casado-Vara *et al.*, (2018)	Hua *et al.* (2018)	Galvez *et al.*(2018)	Kshetri (2018)	Leong *et al.*(2018)	Lin *et al.* (2018)	Perboli *et al.* (2018)	Rejeb (2018)
Benefits	Supply chain and traceability efficiency	✓	✓	✓	✓	✓	✓	✓	✓	✓	✓	✓	✓		✓	✓	✓	✓	✓	✓	✓
	Information transparency and supply chain visibility	✓	✓	✓	✓	✓		✓	✓	✓	✓	✓		✓	✓	✓	✓	✓	✓	✓	✓
	Information authenticity and accountability	✓	✓	✓	✓	✓	✓		✓	✓	✓	✓	✓	✓	✓	✓	✓	✓		✓	✓
	Supply chain digitisation			✓		✓	✓	✓	✓							✓	✓		✓		
	Supply chain resilience									✓						✓	✓		✓	✓	
	Sustainability	✓			✓	✓		✓	✓			✓				✓	✓	✓		✓	✓
Intra-organisational barriers	Company determination and readiness				✓												✓				
	Company demand and understanding				✓		✓									✓		✓			
	Financial constraints	✓			✓			✓													
Inter-organisational barriers	Supplier engagement				✓											✓	✓	✓		✓	
	Supplier financial pressure																				
	Privacy concerns				✓			✓							✓	✓	✓	✓	✓		
	Consumer engagement																				
Systems related barriers	Scalability and storage	✓			✓					✓		✓			✓		✓			✓	
	Lack of standard																				
	Raw data manipulation				✓			✓									✓		✓		
External barriers	Lack of governmental support																				✓
	Policy deficiency				✓			✓									✓				✓

Categories	Papers	Sander *et al.*, (2018)	Tan *et al.*, (2018)	Thiruchelvam at al., (2018)	Tripoli and Schmidhuber (2018)	Verhoeven *et al.*, (2018)	Yiannas (2018)	Antonucci *et al.*, (2019)	Kamilaris *et al.*, (2019)	Mondal *et al.*, (2019)	Pearson *et al.*, (2019)	Salah *et al.*, (2019)	Van Hoek (2019)	Zhao *et al.*, (2019)	Al-Amin *et al.*, (2020)	Alkhudary *et al.*, (2020)	Awan *et al.*, (2020)	Bumblauskas *et al.*, (2020)	Casino *et al.*, (2020)	Cointelegraph, VeChain, (2020)	Dutta *et al.*, (2020)
Benefits	Supply chain and traceability efficiency	✓	✓	✓	✓	✓	✓	✓	✓		✓		✓	✓	✓	✓		✓	✓	✓	✓
	Information transparency and supply chain visibility	✓	✓	✓	✓	✓	✓	✓	✓		✓		✓	✓	✓	✓		✓	✓	✓	✓
	Information authenticity and accountability	✓			✓	✓	✓		✓		✓	✓		✓					✓		✓
	Supply chain digitisation	✓	✓		✓		✓							✓	✓			✓	✓	✓	
	Supply chain resilience	✓	✓	✓	✓	✓	✓			✓				✓	✓		✓		✓	✓	✓
	Sustainability				✓				✓		✓		✓	✓					✓	✓	✓
Intra-organisational barriers	Company determination and readiness			✓	✓	✓						✓									✓
	Company demand and understanding	✓		✓					✓					✓	✓						✓
	Financial constraints			✓	✓						✓	✓		✓		✓			✓		✓
Inter-organisational barriers	Supplier engagement	✓							✓		✓	✓	✓								
	Supplier financial pressure										✓			✓							
	Privacy concerns	✓			✓				✓		✓			✓							
	Consumer engagement								✓		✓	✓		✓					✓		✓
Systems related barriers	Scalability and storage		✓		✓				✓		✓		✓	✓					✓	✓	✓
	Lack of standard							✓	✓		✓		✓	✓		✓		✓	✓	✓	✓
	Raw data manipulation													✓			✓		✓	✓	✓
External barriers	Lack of governmental support	✓			✓									✓				✓	✓	✓	
	Policy deficiency	✓			✓							✓									✓

(continued)

Table 2.5 (continued)

Categories	Papers	Feng *et al*., (2020)	Howson, (2020)	Iftekhar.*et al*., (2020)	Kayikci *et al*., (2020)	Kohler and Pizzol (2020)	Lin *et al*., (2020)	Novianti *et al*., (2020)	Pournader *et al*., (2020)	Rogerson and Parry (2020)	Skender and Zaninovic (2020)	Song *et al*., (2020)	Tan *et al*., (2020)	Tsolakis *et al*., (2020)	van Hilten *et al*., (2020)	Ali *et al*., (2021)	Kouhizadeh *et al*., (2021)	Saurabh and Dey (2021)	Stranieri *et al*., (2021)
Benefits	Supply chain and traceability efficiency	✓	✓			✓	✓		✓	✓		✓	✓		✓	✓		✓	✓
	Information transparency and supply chain visibility	✓	✓	✓	✓	✓	✓	✓	✓	✓	✓	✓	✓	✓	✓	✓		✓	✓
	Information authenticity and accountability	✓				✓	✓	✓	✓				✓	✓	✓	✓		✓	
	Supply chain digitisation					✓							✓	✓	✓	✓			
	Supply chain resilience	✓				✓			✓			✓			✓	✓		✓	
	Sustainability	✓	✓							✓	✓	✓		✓		✓	✓	✓	
Intra-organisational barriers	Company determination and readiness						✓			✓		✓			✓				
	Company demand and understanding									✓									
	Financial constraints	✓			✓	✓	✓		✓	✓		✓	✓		✓	✓	✓	✓	
Inter-organisational barriers	Supplier engagement			✓	✓	✓			✓	✓						✓	✓	✓	
	Supplier financial pressure		✓		✓	✓			✓	✓					✓	✓	✓		
	Privacy concerns								✓	✓							✓		
	Consumer engagement	✓				✓	✓	✓	✓				✓		✓				
Systems related barriers	Scalability and storage	✓	✓							✓	✓	✓							
	Lack of standard	✓	✓			✓	✓		✓	✓		✓	✓		✓	✓	✓		✓
	Raw data manipulation	✓				✓			✓			✓			✓				
External barriers	Lack of governmental support					✓							✓		✓		✓		
	Policy deficiency	✓	✓	✓	✓		✓		✓	✓							✓		

Table 2.6 The comparisons between SCPV, PBV, RV, and RBV (Adapted from Bromer et al., 2019)

	SCPV	PBV	RV	RBV
Organisational level of analysis	Inter-organisational	Intra-organisational	Inter-organisational	Intra-organisational
Explaining variable (cause)	Practices (imitable and accessible)		Resources (rare, impossible to imitate or substitute)	
Dependent variable (effect)	Business and relational performance	Unit business performance improvement	Competitive advantage	
Isolating mechanisms	Weak or non-existent isolating mechanisms		Extremely strong isolating mechanisms	

stakeholder cooperation, (iv) trade secrets, and (v) raw data manipulation. By listing and dividing papers, as in Table 2.6, the brief content and highlights of every paper can be shown more clearly.

2.4.2 *Thematic Analysis*

Many studies pointed out the problems in the current food supply chain, including inefficient traceability, information asymmetry, information fraud, and poor supply chain management, among other aspects. In this case, blockchain is supposed to be a feasible solution for the problems mentioned above. The following is the summary of the 58 papers and an initial framework (Fig. 2.2).

2.4.2.1 Benefits

1. Supply chain and traceability efficiency
   Blockchain can improve supply chain efficiency (Ali et al., 2021). Different food companies have different traceability systems, either by paper or by computer systems. However, due to the complex features of the food products, the food supply chain can be very complicated with multiple suppliers both horizontally and vertically (Pearson et al., 2019). The complex food supply chain also means higher risks of recall (mislabelling, chemical contamination, low quality of raw material, food additives) and higher requirements on stakeholders' collaboration. A single-ingredient-caused food recall can take up to months (Lin et al., 2018; Yiannas, 2018). According to Yiannas (2018), it can cost up to $93 billion to recall products due to the inability to trace the root cause of the outbreak. It is believed, therefore, that Blockchain as a promising technology can improve traceability efficiency and enhance trust during food recall (Yiannas, 2018).

   A pilot study of the blockchain-based traceability system was undertaken in 2016 by Walmart and IBM. The two companies worked together to trace

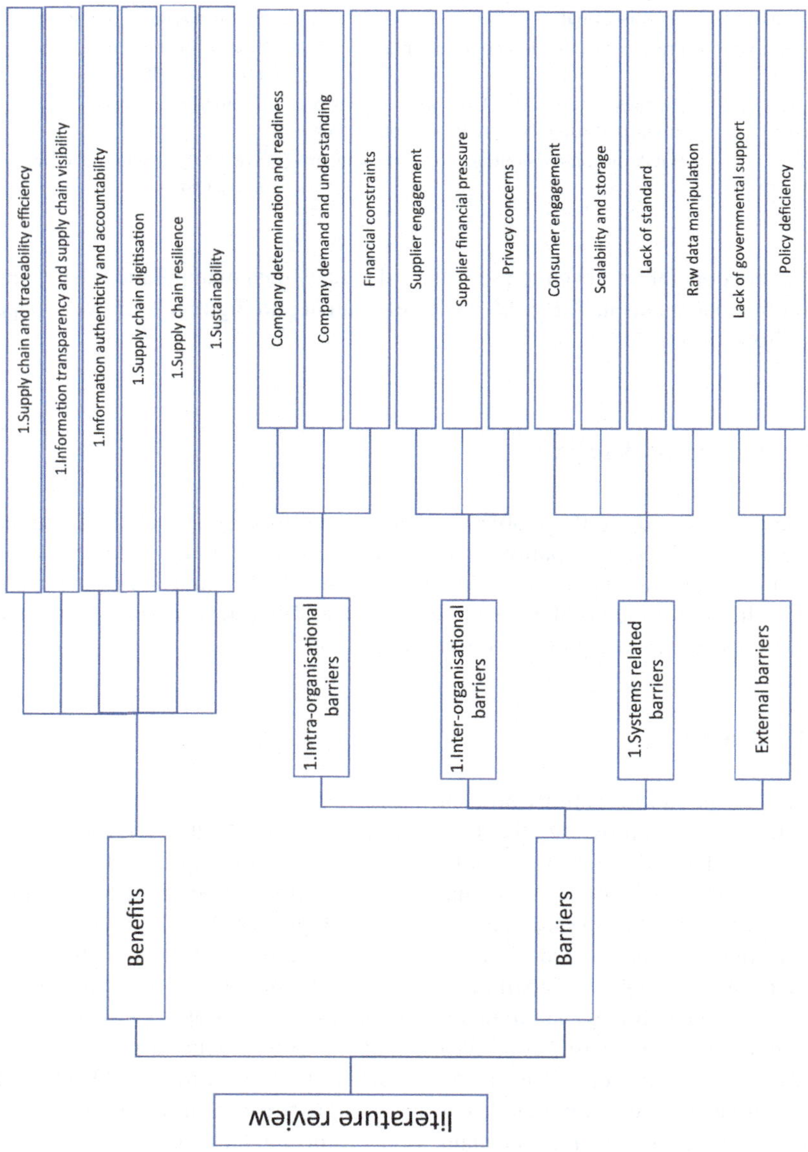

Fig. 2.2 Benefits and barriers of the application of blockchain in food supply chain management: an initial framework

mangoes from a farm to the store. By the current traceability system, it took almost 7 days to collect all the information on the mango movements, which required every stakeholder to contact each other to get to know the required details (Yiannas, 2018). By blockchain, the time to contact and wait for a response from other stakeholders can be eliminated. The movements of mangoes are recorded by each stakeholder along the supply chain, and ready to be checked at any time. The trace time reduced from nearly 7 days to 2.2 s by using blockchain (Yiannas, 2018). AgriDigital and CBH group also carried out a pilot study in the Australian grain industry and found that the blockchain network allowed better traceability efficiency.

2. Information transparency and supply chain visibility

   In the current supply chain, big food brands can normally choose to release partially selected information to the public and aim to benefit the companies themselves, which can mean that customers do not have enough knowledge of products and companies may not be familiar with their suppliers. For certain products that have specific requirements such as halal food, transparency of the supply chain is significantly necessary to ensure products' quality and retain customer trust (Ali et al., 2021; Tieman & Darun, 2017). Although governments and authorities have published policies and make regular checks on food quality in most cases, the authority power can be the target of bribery, and contribute to covering up for big brand companies. For example, the Sanlu milk scandal did not get exposed in the first place because the company managers and local authorities hid the scandal (Barboza, 2008). Information credibility can be a serious issue in centralised food supply chains (Hua et al., 2018). One of the main features of blockchain is decentralisation, which allows authorised users to make transactions and to access history directly without central power intervention. Every registered valid user has the same power to examine a transaction and have a copy of the history. This feature can eliminate any large powers over the information flow, address information asymmetry between stakeholders, and provide transparency along the supply chain. Thus, information visibility is considered as one of the key benefits of blockchain (van Hoek, 2019).

3. Information authenticity and accountability

   Information fraud is another major concern in the current supply chain. van Hilten et al. (2020) suggested that data accountability is one of the drivers for a company to adopt blockchain. Trust is hard to achieve by a centralised supply chain where transparency and visibility remain low. Although companies can release certain information based on requirements, it is easy to change information or erase history to escape from responsibilities or to hide the truth (Biswas et al., 2017; Caro et al., 2018; Tian, 2016, 2017). To address this, once data are updated onto the blockchain, the recordings become permanent. This immutability feature can be achieved by running the blockchain mining process (Tian, 2017). Once the majority of miners/users vote to validate certain transactions, the transaction data remain stored and can never be changed without notifying all the users (Tian, 2017). The history of the products' movements in a supply chain, therefore, can be retrieved and checked at any time whenever needed without worrying that it may have been tampered with. For food products, verifications are necessary to prove companies' eligibility for producing or selling.

4. Supply chain digitisation

   Many studies have mentioned the integration of blockchain with other information systems such as IoT, to digitise the supply chain and to improve efficiency (Cole et al., 2019; Galvez et al., 2018; Leong et al., 2018; Lin et al., 2018; Queiroz et al., 2019; Tian, 2016; Tian, 2017; Tsolakis et al., 2020). Among all the other information systems, the integration based on blockchain and IoT is a hot trend, which includes RFID (Radio-Frequency Identification), GPS (Global Positioning system), GIS (Geographic Information System), WSN (Wireless sensor network), etc. IoT is an intelligent, reliable, and high-speed information network that connects objects, which means that information such as temperature and humidity can be captured automatically by IoT sensors. This real-time information-capturing ability is particularly important for frozen and fresh food products, as the quality is closely related to the external environment (Lin et al., 2017; Tian, 2016). Automation by IoT can increase the efficiency of monitoring and capturing information and reduce human intervention and manual errors (Lin et al., 2018; Tian, 2016, 2017). By combining it with smart contract, once anything goes wrong such as losing temperature control, the digitalised program can be triggered automatically and send registered users' warnings, which can prevent further damage (Caro et al., 2018; Lin et al., 2017; Tian, 2017).

   The Accenture Traceability Report listed a few blockchain pilot studies that have incorporated IoT, such as the WWF using smart tagging combined with blockchain to prevent illegal tuna fishing in Fiji; and Belagricola using IoT and smart contact to track grains and ensure the quality (Leong et al., 2018). Tian (2016, 2017) built an agri-food traceability system based on RFID tags and blockchain to deliver real-time information of food products. From production to retailing, the integration of the two technologies allows data collection and transferring without human intervention and ensures food quality and safety. Lin et al. (2018) proposed a blockchain and IoT-based agriculture system and agreed that the new system is trusted and self-organised without human intervention. The paper by Cole et al. (2019, p. 476) also introduced some electronic supply chain management technologies and suggested that blockchain and other information systems "can be combined and are complementary".

   Apart from combining with IoT devices and digitising data collection process, certain certifications can also be digitalised and updated by blockchain, which allows permissioned users access at any time (Yiannas, 2018). The digital records and documents not only can save time over running a manual paper check, but also eliminate risks from manipulation, counterfeiting, and human errors (CBH Group and AgriDigital, 2017). In 2016, Walmart and Tsinghua University tracked pork in China from farm to fork, and digitised certificates on blockchain.

5. Supply chain resilience

   Galvez et al. (2018), who analysed the potential of its uses in food traceability and authenticity, also suggested that blockchain is a powerful tool to reduce supply chain risks. Traditionally, paper records were used within a supply chain.

However, manual data recording can be very time-consuming or easily cause errors. Thus, blockchain can improve supply chain resilience. With smart contract and IoT devices, blockchain not only can digitise data to improve efficiency but also can eliminate possibilities of wrong data being uploaded on blockchain (Lin et al., 2018; Tsolakis et al., 2020). Caro et al. (2018) built two traceability systems based on a combination of blockchain and IoT on Ethereum and Hyperleger Sawtooth, respectively, and confirmed the ability of blockchain on providing data transparency and auditability. To ensure information safety after being uploaded on blockchain, on one hand, decentralisation can enhance network security and reduce the risks of hacking (Casino et al., 2020). On the other hand, the immutability feature can ensure information authenticity (Lin et al., 2017).

6. Sustainability

   By using blockchain, the food supply chain is found to be more sustainable by operating a more efficient and targeted food recall. When products' information is updated on blockchain in near real-time speed, stakeholders can have more knowledge of products' flow and react to situations quicker. For instance, Walmart realised that fresh import products such as mangoes can wait to be checked up to 4 days at the country's border (Yiannas, 2018). In this case, Walmart can follow up the products' movements, accelerate the products' checking process, and give products longer shelf-lives. The improvement of information transparency can improve supply chain efficiency and eliminate unnecessary product waste.

   Meanwhile, untargeted food recall is also a major cause of food wastes. Food products tend to be complicated with mixtures of many ingredients. Single-ingredient-contamination-caused food recall can be complicated and time-consuming, which requires the traceability system to be extremely efficient. By blockchain-based traceability, it is possible to retrieve needed information, isolate products from certain suppliers, and narrow down the products' recall range. Meanwhile, depending on the information, companies can also make more accurate customer demand forecasting based on the point of sales data (Wang et al., 2019a, b).

2.4.2.2 Barriers

All in all, blockchain sounds like a promising technology that may revolutionise food supply chain, improve efficiency and eliminate risks. More pilot studies have been tested and received positive feedback (CBH Group and AgriDigital, 2017; Yiannas, 2018). However, it is undeniable that blockchain is still at its infancy stage and has a long way to go before it can be widely put in use. There are four categories of barriers that need to be addressed according to Saberi et al. (2018): these are intra-organisational barriers, inter-organisational barriers, systems-related barriers, and external barriers.

1. Intra-organisational barriers

   The intra-organisational barriers are those present within the organisation, such as lack of readiness to adopt blockchain, lack of deep understanding of the technology, and financial constraints. Studies show that many people working on supply chain management are still having trouble in fully understanding the potential of blockchain (Galvez et al., 2018; Hackius & Petersen, 2017; Queiroz et al., 2019; van Hoek, 2019; Zhao et al., 2019). The level of understanding of the technology can significantly impact the participants' attitudes. According to the survey by Hackius and Petersen (2017), participants who are more familiar with the concept and more experienced tend to hold a more positive attitude to blockchain adoption. After analysing some of the most recent pilot studies, Verhoeven et al. (2018) also suggested that there is still the lack of deep understanding of blockchain potentials as many companies tended to choose blockchain as a solution before diagnosing the company's issues. For instance, Verhoeven et al. (2018) suggested that in Walmart's pilot study of tracing mangoes, the tracing speed increase by blockchain should be due to the eliminating of the manual validation process rather than changing to an efficient platform. The mango pilot study also failed to present the importance of the immutability feature, as the records can be very important for fresh food quality. For example, immutable records including temperature and humidity can be evidence for keeping the food value chain. Leong et al. (2018) also suggested that different stages of the supply chain might have different requirements on technology adoption. Therefore, to give the right "remedy" to company problems, it is always necessary to have a comprehensive understanding of both costs and benefits of the technology; sometimes, though, other existing technologies can be better solutions (Leong et al., 2018; Verhoeven et al., 2018).

   Financial constraint is another barrier for some companies, particularly for SMEs. The blockchain implementing and infrastructure maintaining fee can be barriers for them to adopt new innovations (Ali et al., 2021; Leong et al., 2018; Pearson et al., 2019; Perboli et al., 2018; Pournader et al., 2020; Tan et al., 2018). Therefore, it is important to make blockchain "SME friendly" which means easy to use, and easy to deploy with low initial costs (Leong et al., 2018; Pearson et al., 2019). Perboli et al. (2018) developed a blockchain model based on Hyperledger Fabric for SMEs and came to the conclusion that the implementing fee of blockchain can be highly sustainable and can be paid back by saving on costs. They also suggested that replacing the system partially by blockchain is more reasonable than replacing the whole system.

2. Inter-organisational barriers

   The inter-organisational barriers relate to the cooperation between stakeholders within a supply chain. Rather than paper recording, blockchain requires more than one stakeholder to get involved, from raw material suppliers to customers along the multiple tiers of the supply chain. Bringing all stakeholders on board can be an obstacle due to the different levels of awareness and infrastructures. Thus, stakeholder acceptance is one of the barriers at inter-organisational level (Tan et al., 2018). To encourage suppliers to accept blockchain, supplier education about blockchain knowledge and benefits is necessary (Tan et al., 2018). Moreover, financial pressure and privacy concerns are also key challenges. Financial pressure can limit the number of stakeholders adopting blockchain. Some suppliers may have concerns over investment and replacing infrastructure. In this case, apart from making blockchain cheaper to adopt, leadership of the lead company can also encourage stakeholders to join the blockchain project. Van Hoek (2019) analysed three case studies and suggested that blockchain implementation is positively relates to leadership commitment. Ali et al. (2021) also found that the engagement of senior managers, particularly the owner of the firm, is one of the influencing factors to blockchain implementation.

While some companies' databases can be too private to be useful for other stakeholders, which can cause a lack of stakeholder cooperation or lack of supervision within a supply chain (Hua et al., 2018). Therefore, the privacy issue is another concern for stakeholders to join the blockchain platform (Kamilaris et al., 2019; Kouhizadeh et al., 2021; Pearson et al., 2019; Tripoli & Schmidhuber, 2018; Van Hilten et al., 2020; Zhao et al., 2019). The uploaded information can be accessed by any users on the chain, no matter whether it is a permissioned or permission-less blockchain. There is the possibility for disclosure of trade secrets, such as information relating to costs. Van Hilten et al. (2020) found that many supply chain partners worried about losing competitiveness by sharing information so hesitated to do so. In this case, the regulations on data privacy are necessary to protect company data (Leong et al., 2018; Pearson et al., 2019).

Suppliers can register themselves as authorised users which will allow them to upload information, to verify transactions and to access past records. Customers as the end users will also have rights to require about and check products' history. In this case, consumer engagement is recognised as one of the barriers (Rogerson & Parry, 2020). Developing blockchain to be consumer-friendly and smart phone-friendly can potentially increase the usability of the blockchain platform (Rogerson & Parry, 2020).

3. Systems-related barriers

System-related barriers are the barriers related to the technology itself, such as scalability issues, raw data manipulation, and lack of common standard. Although blockchain can provide a robust way to keep records, many studies expressed concerns about raw data manipulation, for example by tampering with IoT sensors, it is hard to know if the raw data in the first place are authentic (Galvez et al., 2018; Lin et al., 2017). It is also possible to damage products on purpose without notifying blockchain users (Kshetri, 2018). Targeting potential raw data manipulation, third parties such as governments and certifications can get involved in the blockchain network by making regular checks to ensure raw data authenticity (Leong et al., 2018; Tian, 2017). Meanwhile, the immutable recording can be used as strategic tool to encourage suppliers to take responsibilities for their products and provide authentic information in the first place.

Another barrier is blockchain scalability, which is also called the "scalability trilemma" by the founder of Ethereum—Vitalik Buterin (Perboli et al., 2018). According to him, it is hard to achieve decentralisation, scalability, and security at the same time—only two out of three can be achieved at one time (Ometoruwa, 2018; Perboli et al., 2018). Bitcoin, for example, was designed to be decentralised and secure with a promising scalability. Scalability determines how large the capacity of the network can be. Currently, the smart contract platform, Ethereum, for example, can process 15 transactions per second, while other platforms such as Visa can process 45,000 transactions per second (Coindesk, 2019). By operating a complicated mining process to validate transactions and keep all transaction copies in each node, blockchain can achieve a high degree of decentralisation and security; however, it can also cause a slow speed of validations particularly when a large number of transactions are happening. This means that high scalability can increase security risks, while low scalability can cause transaction crowds and slow down the network.

Food supply chains tend to be vast with a large number of users existing in one transaction, the global food supply chain scale can reach about Petabytes per year by assumption according to Pearson et al. (2019). Therefore, developers are still working to find a better solution to expend blockchain scalability while maintaining high security and decentralisation. Due to the scalability issues, Pearson et al. (2019) assumed that blockchain is more likely to happen in niche areas in a food supply chain, where blockchain potentials are necessarily needed. Leong et al. (2018) also suggested that the different stages of the food supply chain may have different requirements for blockchain adoption, where the balance of the three features can be different.

Lack of common standard refers to the issue of governance. Rogerson and Parry (2020) found that the data on blockchain can be entered inconsistently. How much data and what kind of data should be captured and uploaded should be addressed. Van Hoek (2019) and Tan et al. (2020) also suggested that the difference of systems and blockchain can lead to data integrity barriers. Thus, to solve the standard issue, Tan et al. (2020) suggested that companies can use the existing standard for blockchain such as the data standard developed by GS1 or develop new standards from scratch such as the Ethereum blockchain platform.

4. External barriers

The external barriers refer to the attitude of governments and external parties. As blockchain is an open database, policies will be needed to protect users' rights and trading secrets. Tse et al. (2017) applied PEST analysis to examine the uncontrollable external environment which includes political, economic, social, and technological factors for blockchain implementation. The paper suggested that governments can obtain supply chain information and reduce food risks by using blockchain. A few countries and authorities also showed their interest in and support for blockchain development (Tse et al., 2017): China, for example, has published the Blockchain White Book and launched blockchain-related projects (Tse et al., 2017).

Although ISO Blockchain (TC307) was working on developing global blockchain standards (Pearson et al., 2019), there is still no strict blockchain policy in food supply chain area to date. Leong et al. (2018) and Pearson et al. (2019) suggested that policies and rules need to be developed to protect users, including what data should be uploaded, who own the data, how to use and store the data, and so on. Therefore, from protecting users' rights point of the view, it is difficult to invite all companies or people to use blockchain before some completed policies are launched.

2.5 Theoretical Basis

The innovation process model and practice-based view (PBV) are applied in this research as the lens to explore how blockchain application was adopted by food companies. The innovation process model involves four steps that deconstruct how companies decided to adopt blockchain to solve certain food supply chain issues. The PBV, on top of the innovation process model, can provide more details of the blockchain implementation process, including influencing factors, barriers, and benefits. Thus, the following sections introduce the innovation process model and PBV, respectively, and explain the reasons and benefits of adopting the two theories.

2.5.1 *Innovation Process Model*

The innovation theory is proposed by Rogers (1995) and explains the process of how "an innovation spreads via communication channels over time among the members of a social system" (Roger et al., 2005, p. 3). The theory suggests the innovation diffusion process by four key influencing factors (innovation itself, communication channels, time, and a social system) and five adopter categories (innovators, early adopters, early majority, late majority, and laggards).

Many researchers emphasise the importance of the innovation process in innovation management (Bessant, 2004; Flint et al., 2005; Sawhney et al., 2006; Su et al., 2011). The question of how the companies organise and conduct the innovation process is an important aspect to understand innovation. For example, the logistics innovation process model was first proposed by Flint et al. (2005) to discover how logistics companies adopt logistics service innovation.

Rogers (1995, p. 11) defined innovation as "an idea, practice, or object that is perceived as new by an individual or other unit of adoption". Innovation is more than a new technology, but can be something new on the products, services, processes, positions, and business models (Flint et al., 2005; Sawhney et al., 2006; Schumpeter, 1934; Su et al., 2011). Innovation is not only new to the world but also can be new to the users (Ageron et al., 2013; Flint et al., 2005; Su et al., 2011).

Innovation in supply chain management usually relates to inter-organisational practices, as supply chain management is an integration of the business process (Ageron et al., 2013; Lambert & Cooper, 2000). Supply chain practices are the activities to manage supply chains and play an important role in supply chain management (Silva et al., 2018). Innovative supply chain practice is defined as "a set of methods and tools that are previously inexistent in companies and/or their subsidiaries that will be generated, developed and deployed within supply chains to tackle different supply chain issues such as quality, costs, and lead-time" (Ageron et al., 2013, p. 267).

The business process is defined by Ljungberg and Larsson (2001, p. 44) as "a repetitively used network of orderly linked activities using information and resources for transforming 'object in' into 'object out', extending from the point of identification to that of the satisfaction of the customer's needs". Moreover, the process orientation is "the acceptance of the process definition and organization-wide efforts in its adoption" (Davenport, 1992; Flint et al., 2005, p. 120). Based on the definitions of innovation and process, the innovation process model mainly explains the "how" question (Su et al., 2011). Different from innovation diffusion theory, which focuses on innovation in the later stages, the innovation process investigates idea generation and the following activities. It focuses on the question of "how" companies innovative.

The model was first proposed under the service provider context by Flint et al. (2005). Flint and colleagues used the grounded theory approach and conducted interviews with senior logistics managers to develop an innovation process model with four processes (Grawe, 2009). The model was adopted and expanded by several researchers. Bjorklund and Forslund (2018) proposed a sustainable logistics innovation (SLI) process model (idea generation, selection of ideas, concept of development, making sustainability business case, implementation, and learning) based on the original model. The finding supports the importance of managing all phases and formalisation of the innovation process. The research also points out the critical activities, challenges, and key actors in each phase. de Mota Pedrosa et al. (2015) adopted a logistics innovation development model (idea generation, concept development, business analysis, implementation) to investigate the employees' behaviours with customers. Wagner (2008) proposed five innovation activities (internal research and development, external research and development, investment in infrastructure and capital goods, acquisition of knowledge, training, and further education) in the German transportation industry.

The innovation process model was also adopted and enriched by Su et al. (2011) in the context of an innovative hospital supply chain. Su et al. (2011) suggested that the term "supply chain innovation process" can also be used when multiple organisations were involved in the case study. This means that the model is not strictly limited to logistics but can also be used in supply chain management. In addition, the model by Flint et al. (2005) was created and discussed from the customer value-oriented perspective. Su et al. (2011) proposed that the model is also valid when involving suppliers.

To be more specific, the logistics innovation process includes four stages:

Setting the stage activities: The first stage is to create an interactive environment to lay a foundation for the next stage. This stage includes activities such as gathering key partners and useful resources, discussing the issues, and identifying demands. For example, in the case study by Flint et al. (2005), case company managers hired professionals in this stage to gather customer information in the next stage. Bjorklund and Forslund (2018) suggest that this is a proactive stage, which includes identifying issues and generating ideas. At this stage, the ideas are generated informally by a few key internal actors. This stage focuses on all kinds of preparations—as the saying goes: "a workman must first sharpen his tools if he is to do his work well".

Customer clue-gathering activities: The second stage is about gaining deep insights of the outcomes from the last stage. The customer clue-gathering stage focuses on all kinds of data collection, such as stakeholder perceptions, industry information, technology information, and others. In the case by Flint et al. (2005), a few data collection methods were used by the case companies, such as conducting interviews to understand customers' value perceptions, reading trade journals, and attending industry conferences. This is also a stage to analyse and select quality ideas and available resources from the first stage (Bjorklund & Forslund, 2018). Moreover, the research by Su et al. (2011) verified that this stage can not only be used in the customer value context but can also be used for suppliers in the supply chain innovation context. Sufficient knowledge can help companies to further clarify opportunities of improvements, which is essential to successful innovation (Chapman et al., 2002). Wagner (2008) suggested that this stage requires industry expertise as customers may not always have enough knowledge.

Negotiating, clarifying, and reflecting activities: Under the context of logistics provider, the model by Flint et al. (2005) suggested that this stage is about internal repetitive discussions and idea exchanging on the data collected from the last stage. However, Su et al. (2011) suggested that continuous discussions and data collection can happen both internally and externally. It also includes negotiation with suppliers to persuade them to accept the innovations. Moreover, Su et al. (2011) extended this stage to innovation design based on the information from the first two stages. This stage is summarised as a concept development and business case making stage by Bjorklund and Forslund (2018) under the sustainable logistics innovation context. To develop ideas to an implementable concept, this stage includes activities of deepening knowledge, analysing potential opportunities and risks, and valuating consequences of the innovation. This stage also stresses the importance of knowledge sharing and information sharing between supply chain partners.

Inter-organisational learning: The final stage is an implementation stage that involves joint learning by companies and stakeholders (Bjorklund & Forslund, 2018). This stage is about the final launch innovation and gaining new understandings and insights, as well as exploring possibilities and potentials from the innovation. Therefore, this stage emphasises cross-organisational activities and actors. Similar to the last stage, this is also an idea-exchanging phase but a larger scale of stakeholders are involved. From the stakeholder perspective, the activities should engage more stakeholders and understand their opinions such as providing workshops and training (Flint et al., 2005). From the innovation perspective, the innovation should also be updated constantly from the initial innovation design stage to become an innovation process loop. For supply chain innovation, Su et al. (2011) included all supply chain participants to jointly implement the new system.

The four stages present the process of innovative idea generation and decision-making. The research by Su et al. (2011) also examined the performance of innovation process by comparing the financial results, the changes in the supply chain, and the improvement in supplier relationships.

2.5.2 *Practice-Based View (PBV)*

Practice is defined as "a way of doing that is embedded in the context of inwardly and outwardly interlinked elements that focus on performance" (Bromiley & Rau, 2014, p. 532; Korkman, 2006). It is more than an action, a process, or a result, but involves actors, actions, process, context, and tools, among other elements. Practice is the results of the world and the process; it provides a way to understand "knowledge in-action" (Gherardi, 2000; Dougherty, 2004). Korkman et al. (2010) claimed that practices can integrate resources and create values (Russo-Spena & Mele, 2012).

The PBV was proposed by Bromiley and Rau (2014) to understand the determinants of company behaviours and performance. From a practice-based lens, the knowledge can be captured from the ongoing practice and give specific and feasible action advice to managers and practitioners, including what to do and what not to do (Dougherty, 2004). This is because the PBV theoretical lens can present all kinds of performance implications of practices, including negative performance such as issues (Bromiley & Rau, 2014, 2016a). Thus, many researchers suggest that the PBV should be applied to study performance-based research, such as the impact of supply chain technologies (Bromiley & Rau, 2014; Carter et al., 2017; Lin et al., 2016; Treacy et al., 2019).

The PBV is deeply connected with Behavioural Theory of the Firm (BTOF). BTOF considers organisations as *routines* (Cyert & March, 1963). Thus, the impact of practices is not only influenced by practices themselves, but also depends on the company management and parameters, and the interactions with others (Bromiley & Rau, 2014). The PBV also connects with evolutionary economics, which stresses the importance of both individual and society (Winter, 1964, 1994). It also suggests that practice can be influenced by the context and firm's history. Besides, practices can strongly impact the organisation's capabilities, which suggests that PBV relates to the company's capabilities (Bromiley & Rau, 2014).

Many researchers compared the PBV with the Resource-based View (RBV) (Bromiley & Rau, 2014, 2016a; Bromer et al., 2019; Carter et al., 2017) (Table 2.6). However, different from the RBV, which suggests that unique resources can bring competitiveness, the PBV examines the performances from imitable and transferable activities across different companies (Bromiley & Rau, 2014, 2016b). Bromiley and Rau (2014) proposed that the dependent variable and isolating mechanisms are the two differences between RBV and PBV. In contrast with the RBV, the PBV emphasises the practices that can be accessible for other companies and focuses on performance improvements (Treacy et al., 2019).

A few researchers discussed that the PBV and RBV should not substitute each other but complement each other (Almohri, 2018; Carter et al., 2017; Hitt et al., 2016; Silva et al., 2018). "Practices are often coupled with resources to be implemented in an effective way. The specific interplay between resources and practices may decide about how the practices actually impact organisational and supply chain performance" (Silva et al., 2018, p. 661). In other words, resources should also be involved when analysing certain practices. The PBV can provide an approach to resource-integrating actions and interactions (Bromer et al., 2019; Russo-Spena & Mele, 2012).

Russo-Spena and Mele (2012) adopted the PBV to investigate the value-creating process in innovation. The PBV presents how the resources can be integrated to create values. "The practice-based view perceives innovation as a set of actions that the subjects performing the innovative practice accomplish by using certain resources" (Korkman et al., 2010; Russo-Spena & Mele, 2012, p. 544). Moreover, the research also suggests that the PBV can provide an approach to social connections and contexts (Russo-Spena & Mele, 2012). In other words, the PBV provides more than just the outcomes of innovation; it also helps to analyse innovation from the social connections, the context, and the intermediaries.

Treacy et al. (2019) applied the PBV to examine the impact of transferable ISO14001 environmental practices on company operating performance. The research aligned with Hitt et al. (2016) and suggested that the PBV has real value to study operations management. It is applicable to provide insights of practices and examine the entire range of performance. The PBV emphasises that practices should be imitable and transferable, so they can be applied to various companies in the industry.

Within a supply chain management research, Carter et al. (2017) proposed a Supply Chain Practice View (SCPV) based on the Relational View (RV) and the PBV. Similar to the RBV, the RV also concerns the resources that can bring competitive advantage to companies, but at an inter-organisational level. It spans from a single organisation to various organisations. Thus, the SCPV also considers the imitable and transferable practices, but suggests that there are intra-organisational and inter-organisational practices within a supply chain (Shaw et al., 2021). *Intra*-organisational practices happen within a firm, while *inter*-organisational practices need mutual efforts from at least two firms within a supply chain (Carter et al., 2017; Silva et al., 2018). In other words, collaboration is required in supply chain practices. SCPV is a combination of RV and PBV, which focuses on the inter-organisational level and concerns combined performance across firms. Shaw et al. (2021) adopted the SCPV to identify the impact of implementing environmental supply chain management at the level of intra- and inter-organisational levels through the SCPV theoretical lens. The research suggested that the SCPV can provide the understanding of conditions in which the organisation adopts certain practices, and the impact from both internal and external business environments (Carter et al., 2017; Shaw et al., 2021).

Bromer et al. (2019) adopted both the PBV and the SCPV to investigate sustainability diffusion in the chemical supply chain. The research identified the prerequisites, the internal and external practices, and performance improvements. The findings suggest that the SCPV can help to explain the process of sustainable decision-making (Bromer et al., 2019).

Silva et al. (2018) applied the PBV and the SCPV in the case study of the Brazilian cashew nut supply chain in the case of natural disasters. The PBV provides a way to understand company behaviours and managers' decision-making. The research investigated the cashew companies' supply chain practices in both intra- and inter-levels in responding to natural disasters. The finding suggests that organisational context can impact and stimulate practices, which was also suggested by Bromiley and Rau (2014). Moreover, the research agrees that practices and resources are complementary, and have a joint impact on performance (Carter et al., 2017; Hitt et al., 2016; Silva et al., 2018).

2.6 Research Gaps

After a systematic literature review that summaries the related literatures, a few research gaps are identified and presented in this section.

1. Based on the systematic literature review, most of the research was theoretical and conducted based on assumptions (Kamilaris et al., 2019; Zhao et al., 2019; Pournader et al., 2020), Only a limited number of empirical research studies were conducted in the area of blockchain. As a young technology, empirical research is necessary to help to understand how and why blockchain applications are chosen and implemented in food supply chains. Thus, there is a gap in the literature in this regard. This research aims to fill the gap by conducting empirical research and answering the research questions.
2. As blockchain is in its infancy, many existing blockchain studies relate to the financial industry. Although blockchain is starting to gain traction in the supply chain area, only very limited blockchain research focuses on food supply chain management. Many of the food supply chain papers encompass multiple areas including supply chains, transportation, and logistics (ShipChain, 2018; Pournader et al., 2020). In this case, the lack of literature on the food supply chain is also a gap addressed by this research, which specifically focuses on the food supply chain.

3. The innovation process theory and PBV are relatively new and have only been applied in the supply chain literature to a limited extent. Currently, applications of innovation process theory and PBV in the blockchain area are lacking. Some studies adopted various models in blockchain studies, such as Kamble et al. (2019) adopted TAM (technology acceptance model), TPB (theory of planned behaviour), and TRI (technology readiness index); Francisco and Swanson (2018) adopted UTAUT (unified theory of acceptance and use of technology). However, this research is the first to apply the innovation process theory combined with PBV in blockchain studies.
4. The innovation process theory was proposed with four main stages during a company's decision-making process. Thus, there is an opportunity to refine and enrich the theory while applying it in blockchain studies. Innovation process theory can provide a theoretical lens to the research questions, this research could also add sub-constructs to enrich the theory. PBV examine the determinants to a company's decision-making, which also needs an exploration. By applying PBV in blockchain research, PBV can be refined with a structure.

2.7 Summary

This chapter explores the existing literature about the food supply chain, blockchain, and traceability, and discussed the theoretical lenses that are going to be applied in this research. By conducting a content-based systematic literature review, this chapter summarises six benefits (supply chain and traceability efficiency, information transparency and supply chain visibility, information authenticity and accountability, digitisation, resilience, sustainability) that can be brought by blockchain and four key barriers (intra-organisational, inter-organisational, systems-related, and external barriers) during blockchain implementation. In addition to summarising key knowledge from the existing literature, this chapter also points out some research gaps. For example, there is limited empirical research in this area as most related research is based on theories and assumptions. Another research gap is the lack of applications of the innovation process model and the PBV in blockchain research. Thus, this research aims to fill these gaps and contribute to these areas.

References

Adaptaliftgroup. (2012). *RFID vs Barcodes: Advantages and disadvantages comparison*. Logistics & Materials handling Blog. Accessed April 17, 2019, from https://www.aalhysterforklifts.com.au/index.php/about/blog-post/rfid_vs_barcodes_advantages_and_disadvantages_comparison

Ageron, B., Lavastre, O., & Spalanzani, A. (2013). Innovative supply chain practices: The state of French companies. *Supply Chain Management: An International Journal, 18*(3), 265–276.

Agres, T. (2017). *Blockchain to the rescue: Eliminating global food fraud. Food quality & Safety*. Accessed November 14, 2018, from https://www.foodqualityandsafety.com/article/can-blockchain-technology-eliminate-global-food-fraud/

Aitken, R. (2017). IBM & Walmart Launching Blockchain food safety alliance in China with Fortune 500's JD.com. *Forbes*. Accessed November 11, 2018, from https://www.forbes.com/sites/rogeraitken/2017/12/14/ibm-walmart-launching-blockchain-food-safety-alliance-in-china-with-fortune-500s-jd-com/#ed3616b7d9c5

Al-Amin, S., Sharkar, S. R., Kaiser, M. S., & Biswas, M. (2020). Towards a blockchain based supply chain management for E-Argo business system. In *Proceedings of International Conference on Trends in Computational and Cognitive Engineering*, Vol. 1309, pp. 329–339.

Ali, M. H., Chung, L., Kumar, A., Zailani, S., & Tan, K. H. (2021). A sustainable blockchain framework for the halal food supply chain lessons from Malaysia. *Technological Forecasting and Social Change, 170*, 1–14.

Alkhudary, R., Brusset, X., & Fenies, P. (2020). Blockchain in general management and economics: A systematic literature review. *European Business Review, 32*(4), 765–783.

Almohri, S. M. (2018). *A 'practice-based' theory of the firm: A practice theory perspective to organizational strategy development and knowledge management*. Dissertation, Georgia State University.

Antonucci, F., Figorilli, S., Costa, C., Pallottino, F., Roso, L., & Menesatti, P. (2019). A review on blockchain applications in the agri-food sector. *Journal of the Science of Food and Agriculture, 99*(14), 6129–6138.

Aruoma, O. I. (2006). The impact of food regulation on the food supply chain. *Toxicology, 221*(1), 119–127.

Atlam, H. F., Alenezi, A., Alassafi, M. O., & Wills, G. N. (2018). Blockchain with internet of things: Benefits, challenges, and future directions. *International Journal of Intelligent Systems and Applications, 6*(6), 40–48.

Aung, M., & Chang, Y. (2014). Traceability in a food supply chain: Safety and quality perspectives. *Food Control, 39*, 172–184.

Awan, S. H., Nawaz, A., Ahmed, S., Khattak, H. A., Zaman, K., & Najam, Z. (2020). Blockchain based smart model for agricultural food supply chain. In *2020 International Conference on UK-China Emerging Technologies (UCET)*, Glasgow, UK, pp. 1–5.

Badia-Melis, R., Mishra, P., & Ruiz-Garcia, L. (2015). Food traceability: New trends and recent advances. A review. *Food Control, 57*, 393–401.

Barboza, D. (2008). China milk cover-up started last year. *The New York Times*. Accessed May 22, 2019, from https://www.nytimes.com/2008/09/23/world/asia/23iht-24milk-new.16423573.html

Bermeo-Almeida, O., Cardenas-Rodriguez, M., Samaniego-Cobo, T., Ferruzola-Gomez, E., Cabezas-Cabezas, R., & Bazan-Vera, W. (2018). Blockchain in agriculture: A systematic literature review. In *4th International Conference, CITI 2018, Guayaquil, Ecuador, November 6–9, 2018, Proceedings*, Vol. 883, pp. 44–56.

Bessant, J. (2004). Supply chain learning. In S. New & R. Westbrook (Eds.), *Understanding supply chains* (pp. 165–190). Oxford University Press.

Biswas, K., Muthukkumarasamy, V., & Tan, W. L. (2017). Blockchain based wine supply chain traceability system. In *Proceedings of the Future Technologies Conference (FTC) 2017, Vancouver, BC, Canada*, pp. 56–62.

Bjorklund, M., & Forslund, H. (2018). Exploring the sustainable logistics innovation process. *Industrial Management & Data Systems, 118*(1), 204–217.

Blockgeeks. (2018). What is Ethereum? The most comprehensive guide ever! *Blockgeeks*. Accessed October 30, 2018, from https://blockgeeks.com/guides/what-is-ethereum/

Bosona, T., & Gebresenbet, G. (2013). Food traceability as an integral part of logistics management in food and agricultural supply chain. *Food Control, 33*(1), 32–48.

Bromer, J., Brandenburg, M., & Gold, S. (2019). Transforming chemical supply chains toward sustainability – A practice-based view. *Journal of Cleaner Production, 236*, 1–16.

Bromiley, P., & Rau, D. (2014). Research prospective: Towards a practice-based view of strategy. *Strategic Management Journal, 35*(8).

Bromiley, P., & Rau, D. (2016a). Operations management and the resource based view: Another view. *Journal of Operations Management, 41*, 95–106.

Bromiley, P., & Rau, D. (2016b). Missing the point of the practice-based view. *Strategic Organizations, 14*(3), 260–269.

Bumblauskas, D., Mann, A., Dugan, B., & Rittmer, J. (2020). A blockchain use case in food distribution: Do you know where your food has been? *International Journal of Information Management, 52*, 1–10.

Canavari, M., Centonze, R., Hingley, M., & Spadoni, R. (2010). Traceability as part of competitive strategy in the fruit supply chain. *British Food Journal, 112*(2), 171–186.

Caro, M. P. Ali, M. S., Vecchio, M., & Giareda, R. (2018). *Blockchain-based traceability in agri-food supply chain management: A practical implementation*. IoT Vertical and Tropical Summit on Agriculture – Tuscany (IoT Tuscany), Tuscany, pp. 1–4.

Carter, C. R., Kosmol, T., & Kaufmann, L. (2017). Toward a supply chain practice view. *Journal of Supply Chain Management, 53*(1), 114–122.

Casado-Vara, R. C., Prieto, J., Prieta, F. D. L., & Corchado, J. M. (2018). How blockchain improves the supply chain: Case study alimentary supply chain. *Procedia Computer Science, 134*, 393–398.

Casino, F., Kanakaris, V., Dasaklis, T. K., Moschuris, S., Stachtiaris, S., Pagoni, M., & Rachaniotis, N. P. (2020). Blockchain-based food supply chain traceability: A case study in the dairy sector. *International Journal of Production Research, 2020*, 1–13.

CBH Group and AgriDigital. (2017). *Pilot report: Solving for supply chain inefficiencies and risks with blockchain in agriculture*. Accessed June 26, 2019, from https://www.agridigital.io/reports/blockchain-pilot-report

Chamlee, V. (2016). *Why does the FDA take so long to recall food?* EATER. Accessed November 25, 2018, from https://www.eater.com/2016/8/1/12343626/fda-food-recalls-chuck-schumer

Chapman, R. L., Soosay, C., & Kandampully, J. (2002). Innovation in logistic services and the new business model: A conceptual framework. *Managing Service Quality, 12*(6), 358–371.

Christopher, M. (2016). *Logistics & supply chain management* (5th ed.). Pearson UK.

Christopher, M., Mena, C., Khan, O., & Yurt, O. (2011). Approaches to managing global sourcing risk. *Supply Chain Management: An International Journal, 16*(2), 67–68.

Codex Alimentarius Commission. (2015). *Procedural manual: twenty-fourth edition*. Accessed July 05, 2020, from https://www.fao.org/3/i5079e/i5079e.pdf

Coindesk. (2019). *Will Ethereum scale?* Accessed May 22, 2019, from https://www.coindesk.com/information/willethereum-scale

Cointelegraph and VeChain. (2020). *When blockchain meets IoT: Ensuring food safety in the 2020s*. Accessed February 02, 2021, from https://s3.cointelegraph.com/storage/uploads/view/4612f245e3cf8bc227084dff049f4016.pdf

Cole, R., Stevenson, M., & Aitken, J. (2019). Blockchain technology: Implications for operations and supply chain management. *Supply Chain Management: An International Journal, 24*(4), 469–483.

Cooper, M. C., Lambert, D. M., & Pagh, J. D. (1997). Supply chain management: More than a new name for logistics. *The International Journal of Logistics Management, 8*(1), 1–14.

Costa, C., Antonucci, F., PallottiNo, F., Aguzzi, J., Sarria, D., & Menesatti, P. (2013). A review on agri-food supply chain traceability by means of RFID technology. *Food and Bioprocess Technology, 6*(2), 353–366.

Cousins, P. D., Lamming, R. C., & Bowen, F. E. (2004). The role of risk in environment-related supplier initiatives. *International Journal of Operations & Production Management, 24*(6), 554–565.

Cyert, R. M., & March, J. G. (1963). *A behavioural theory of the firm*. Prentice Hall.

Dabbene, F., Gay, P., & Tortia, C. (2014). Traceability issues in food supply chain management: A review. *Biosystems Engineering, 120*, 1–16.

Dani, S., & Deep, A. (2010). Fragile food supply chains: Reacting to risks. *International Journal of Logistics: Research and Applications, 13*(5), 395–410.

Danovich, T. (2016). How food recalls really work: Companies pull 8,000 products off the shelves each year. Where do they go- and how much does that cost? *Eater*. Accessed April 12, 2019, from https://www.eater.com/2016/1/26/10830614/food-recalls-how-they-work

Davenport, T. H. (1992). *Process innovation: Reengineering work through information technology*. Harvard Business School Press.

Davis, T. (1993). Effective supply chain management. *Sloan Management Review, 34*(4), 35–46.

de Mota Pedrosa, A., Blazevic, V., & Jasmand, C. (2015). Logistics innovation development: A micro-level perspective. *International Journal of Physical Distribution & Logistics Management., 45*(4), 313–332.

Diamond Time-Lapse. (2018). Do you know your diamonds? *Everledger*. Accessed October 27, 2018, from https://diamonds.everledger.io

Dougherty, D. (2004). Organizing practices in services: Capturing practice-based knowledge for innovation. *Strategic Organization, 2*(1), 35–64.

Durach, C. F., Kenbro, J., & Wieland, A. (2017). A new paradigm for systematic literature reviews in supply chain management. *Journal of Supply Chain Management, 53*(4), 67–85.

Dutta, P., Choi, T., Somani, S., & Butala, R. (2020). Blockchain technology in supply chain operations: Applications, challenges and research opportunities. *Transportation Research Part E, 142*, 1–33.

Ene, C. (2013). The relevance of traceability in the food chain. *Economics of Agriculture, 2*(60), 287–297.

EPCglobal. (2009). *The EPCglobal Architecture Framework*. Accessed June 11, 2018, from http://www.gs1.org/sites/default/files/docs/architecture/architecture_1_3-framework-20090319.pdf

European Commission. (2002). Council Regulation (EC) No 178/2002 of the European parliament and of the council. *Official Journal of the European Union, L31*, 1–24.

European Commission. (2019). *Food law general requirements*. Accessed June 02, 2019, from https://ec.europa.eu/food/safety/general_food_law/general_requirements_en

FAO. (2003). *The State of Food insecurity in the world*. Accessed June 02, 2019, from http://www.fao.org/3/j0083e/j0083e00.pdf

Faye, P. S. (2017). Use of blockchain technology in agribusiness: Transparency and monitoring in agricultural trade. In *Advances in economics, business and management research, proceeding of the 2017 international conference on management science*.

Feng, H., Wang, X., Duan, Y., Zhang, J., & Zhang, X. (2020). Applying blockchain technology to improve agri-food traceability: A review of development methods, benefits, and challenges. *Journal of Cleaner Production, 260*, 1–15.

Fink, A. (2005). *Conducting research literature reviews: From paper to the internet*. Sage.

Flint, D. J., Larsson, E., Gammelgaard, B., & Mentzer, J. T. (2005). Logistics innovation: A customer value-oriented social process. *Journal of Business Logistics, 26*(1), 113–147.

Folkerts, H., & Koehorst, H. (1997). Challenges in international food supply chains: Vertical co-ordination in the European agribusiness and food industries. *Supply Chain Management: An International Journal, 2*(1), 11–14.

Francisco, K., & Swanson, D. (2018). The supply chain has no clothes: Technology adoption of blockchain for supply chain transparency. *Logistics, 2*, 1–13.

Franks, J. (2000). Supply chain innovation. *Work Study, 49*(4), 152–155.

Galimberti, A., Mattia, F. D., Losa, A., Bruni, I., Federici, S., Casiraghi, M., Martello, S., & Labra, M. (2013). DNA barcoding as a new tool for food traceability. *Food Research International, 50*(1), 55–63.

Galvão, J., Margeirsson, S., Garate, C., Viðarsson, J., & Oetterer, M. (2010). Traceability system in cod fishing. *Food Control, 21*(10), 1360–1366.

Galvez, J. F., Meijuto, J. C., & Simal-Gandara, J. (2018). Future challenges on the use of blockchain for food traceability analysis. *Trends in Analytical Chemistry, 107*, 222–232.

Gausdal, A. H., Czachorowski, K. V., & Solesvik, M. Z. (2018). Applying Blockchain technology: Evidence from Norwegian companies. *Sustainability, 10*, 6.

Ge, L. Brewster, C., Spek, J., Smeenk, A., & Top, J. (2017). *Blockchain for agriculture and food: Findings from the pilot study*. Wageningen, Technical Report No. 2017-112, Wageningen Economic Research.

Ghemawat, P. (2001). Distance still matters, the hard reality of global expansion. *The Harvard Business Review*. Accessed November 14, 2018, from https://hbr.org/2001/09/distance-still-matters-the-hard-reality-of-global-expansion

Gherardi, S. (2000). *Practice-based theorizing on learning and knowing in organizations* (Vol. 7, pp. 211–223). Sage.

Golan, E., Krisso, B., & Kuchler, F. (2004). *Food traceability: One ingredient in a safe and efficient food supply. Amber waves: The economics of food, farming, natural resources, and rural America*. Accessed December 09, 2019, from https://econpapers.repec.org/article/agsuersaw/129805.htm

Grawe, S. J. (2009). Logistics innovation: A literature-based conceptual framework. *The International Journal of Logistics Management, 20*(3), 360–377.

Grimsdell, K. (1996). The supply chain for fresh vegetables: What it takes to make it work. *Supply Chain Management, 1*(1), 11–14.

Grunert, K. G. (2005). Food quality and safety: Consumer perception and demand. *European Review of Agricultural Economics, 32*(3), 369–391.

Gupta, M. (2017). *Blockchain for dummies*. Wiley.

Hackius, N. Petersen, M. (2017). Blockchain in logistics and supply chain: Trick or treat? In *Digitalization in Supply Chain Management and Logistics*. Accessed May 24, 2019, from https://www.econstor.eu/bitstream/10419/209299/1/hicl-2017-23-003.pdf

Handfield, R. B., & Bechtel, C. (2002). The role of trust and relationship structure in improving supply chain responsiveness. *Industrial Marketing Management, 31*(4), 367–382.

Hitt, M. A., Carbes, C. M., & Xu, K. (2016). A current view of resource based theory in operations management: A response to Bromiley and Rau. *Journal of Operations Management, 41*(1), 107–109.

Howson, P. (2020). Building trust and equity in marine conservation and fisheries supply chain management with blockchain. *Marine Policy, 115*, 1–6.

Hua, J., Wang, X., Kang, M., Wang, H., & Wang, F. (2018). Blockchain based provenance for agricultural products: A distributed platform with duplicated and shared bookkeeping. In *Proceedings of the IEEE intelligent vehicles symposium (IV), Changshu, China*.

IBM. (2017). *Walmart, JD.com, IBM and Tsinghua University Launch a Blockchain food safety Alliance in China*. IBM. Accessed November 14, 2018, from https://www-03.ibm.com/press/us/en/pressrelease/53487.wss

IBM. (2018). *Maersk and IBM introduced TradeLens Blockchain shipping solution*. IBM. Accessed October 22, 2018, from https://newsroom.ibm.com/2018-08-09-Maersk-and-IBM-Introduce-TradeLens-Blockchain-Shipping-Solution

Iftekhar, A., Cui, X., Hassan, M., & Afzal, W. (2020). *Application of blockchain and internet of things to ensure tamper-proof data availability for food safety* (Vol. 2020, pp. 1–14). Wiley.

ISO 8402. (1994). *ISO 8420 1994 ISO Definitions*. Scribd. Accessed October 23, 2018, from www.scribd.com/doc/40047151/ISO-8402-1994-ISO-Definitions

ISO 9000. (2005). Accessed October 23, 2018, from www.pqm-online.com/assets/files/standards/iso_9000-2005.pdf

Jeppsson, A., & Olsson, O. (2017). *Blockchains as a solution for traceability and transparency*. Master's thesis, Lund University, Lund. Accessed October 30, 2018, from https://closer.lindholmen.se/sites/default/files/2021-10/thesis_jeppsson_olsson_2017_0_0.pdf

Kamble, S., Gunasekaran, A., & Arha, H. (2019). Understanding the blockchain technology adoption in supply chains-Indian context. *International Journal of Production Research, 57*(7), 2009–2033.

Kamilaris, A., Fonts, A., & Prenafeta-Boldv, F. X. (2019). The rise of blockchain technology in agriculture and food supply chain. *Trends in Food Science and Technology, 91*, 640–652.

Kayikci, Y., Subramanian, N., Dora, M., & Bhatia, M. S. (2020). Food supply chain in the era of industry 4.0: Blockchain technology implementation opportunities and impediments from the perspective of people, process, performance, and technology. *Production Planning & Control, 33*(2–3), 301–321.

Kelepouris, T., Pramatari, K., & Doukidis, G. (2007). RFID-enable traceability in the food supply chain. *Industrial Management and Data Systems, 107*(2), 183.

Kendall, H., Kuznesof, S., Dean, M., Chan, M., Clark, B., Home, R., Stolz, H., Zhong, Q., Liu, C., Brereton, P., & Frewer, L. (2019). Chinese consumer's attitudes, perceptions and behavioural responses towards food fraud. *Food Control, 95*, 339–351.

Kennedy, Z. C., Stephenson, D. E., Christ, J. F., Pope, T. R., Arey, B. W., Barrett, C. A., & Warner, M. G. (2017). Enhanced anti-counterfeiting measures for additive manufacturing: Coupling lanthanide nanomaterial chemical signatures with blockchain technology. *Journal of Materials Chemistry, 5*(37), 9570–9578.

Kim, C. B., Kwon, S. H., & Park, Y. Y. (2015). The influence of quality management system standards (ISO 9000) on supply chain innovation and business performance. *Journal of Korea Trade, 19*(2), 23–50.

Kohler, S., & Pizzol, M. (2020). Technology assessment of blockchain-based technologies in the food supply chain. *Journal of Cleaner Production, 269*, 1–10.

Korkman, O. (2006). *Customer value formation in practice: a practice theoretical approach.* Swedish School of Economics and Business Administration. No. 155.

Korkman, O., Storbacka, K., & Harald, B. (2010). Practices as markets: Value co-creation in e-invoicing. *Australasian Marketing Journal, 18*, 236–247.

Korpela, K., Hallikas, J., & Dahlberg, T. (2017). Digital supply chain transformation toward Blockchain integration. In *Proceedings of the 50th Hawaii International Conference on System Sciences*, pp. 4182–4191.

Kouhizadeh, M., & Sarkis, J. (2018). Blockchain practices, potentials, and perspectives in greening supply chains. *Sustainability, 10*(10), 1–16.

Kouhizadeh, M., Saberi, S., & Sarkis, J. (2021). Blockchain technology and the sustainable supply chain: Theoretically exploring adoption barriers. *International Journal of Production Economics, 231*, 1–21.

Kshetri, N. (2018). 1 Blockchain's roles in meeting key supply chain management objectives. *International Journal of Information Management, 39*, 80–89.

Kumar, M. V., & Iyengar, N. C. S. N. (2017). A framework for blockchain technology in rice supply chain management. *Advanced Science and Technology Letters, 146*, 125–130.

Kuo, J. C., & Chen, M. C. (2010). Developing an advanced multi-temperature joint distribution system for the food cold chain. *Food Control, 21*(4), 559–566.

La Londe, B. J. (1997). Supply chain management: Myth or reality? *Supply Chain Management Review, 1*, 6–7.

Lambert, D. M., & Cooper, M. C. (2000). Issues in supply chain management. *Industrial Marketing Management, 29*(1), 65–83.

Lambert, D. M., & Enz, M. G. (2017). Issues in supply chain management: Progress and potential. *Industrial Marketing Management, 62*, 1–16.

Leong, C., Viskin, T., & Stewart, R. (2018). *Tracing the supply chain: How Blockchain can enable traceability in the food industry.* Accenture. Accessed December 12, 2019, from https://www.accenture.com/t20190115T192110Z__w__/us-en/_acnmedia/PDF93/Accenture-Tracing-Supply-Chain-Blockchain-Study-PoV.pdf

Lin, Z., Prajog, D., & Oke, A. (2016). Supply chain technologies: Linking adoption, utilization, and performance. *Journal of Supply Chain Management, 52*(4), 22–41.

Lin, I. C., Shin, H., Liu, J.C., & Jie, Y. X. (2017). Food traceability system using blockchain. In *Proceedings of the 79th, IASTEM International Conference, Tokyo, Japan*, pp. 59–64.

Lin, J., Shen, Z., Zhang, A., & Chai, Y. (2018). Blockchain and IoT based food traceability for smart agriculture. In *Proceedings of the 3rd International Conference on Crowd Science and Engineering*, pp. 28–31.

Lin, W., Huang, X., Fang, H., Wang, V., Hua, Y., Wang, J., Yin, H., Yi, D., & Yau, L. (2020). Blockchain technology in current agricultural systems: From techniques to applications. *IEEE Access, 8*.

Lindgreen, A., & Hingley, M. (2003). The impact of food safety and animal welfare policies on supply chain management – The case of the Tesco meat supply chain. *British Food Journal, 105*(6), 328–349.

Ljungberg, A., & Larsson, E. (2001). *Processbaserad verksamhetsutverkling*. Studentlitteratur AB.

Loria, K. (2017). *Walmart invests in food safety for China*. Food Quality and Safety. Accessed November 14, 2018, from https://www.foodqualityandsafety.com/article/walmart-invests-food-safety-china/

Mai, N., Bogason, S. G., Arason, S., Arnason, S. V., & Matthiasson, T. G. (2010). Benefits of traceability in fish supply chains – Case studies. *British Food Journal, 112*(9), 976–1002.

Manos, B., & Manikas, I. (2010). Traceability in the Greek fresh produce sector: Drivers and constraints. *British Food Journal, 112*(6), 640–652.

Mao, D., Wang, F., Hao, Z., & Li, H. (2018). Credit evaluation system based on Blockchain for multiple stakeholders in the food supply chain. *International Journal of Environmental Research and Public Health, 15*(8), 1627.

Mattoli, V., Mazzolai, B., Mondini, A., Zampolli, S., & Dario, P. (2009). Flexible tag datalogger for food logistics. *Procedia Chemistry, 1*(1), 1215–1218.

McMeekin, T. A., Baranyi, J., Bowman, J., Dalgaard, P., Kirk, M., Ross, T., Schmid, S., & Zwietering, M. H. (2006). Information systems in food safety management. *International Journal of Food Microbiology, 112*(3), 181–194.

Mentzer, J. T., DeWitt, W., Keebler, J. S., Min, S., Nix, N. W., Smith, C. D., & Zacharia, Z. G. (2001). Defining supply chain management. *Journal of Business Logistics, 22*(2), 1–25.

Meuwissen, M. O. M., Velthuis, A. G. J., Hogeveen, H., & Huirne, R. B. M. (2003). Traceability and certification in meat supply chains. *Journal of Agribusiness, 21*(2), 167–181.

Mondal, S., Wijewardena, K. P., Karuppuswami, S., Kriti, N., Kumar, D., & Chahal, P. (2019). Blockchain inspired RFID-based information architecture for food supply chain. *IEEE Internet of Things Journal, 6*(3), 5804–5813.

Nagpurkar, A. W., & Jaiswal, S. K. (2015). An overview of WSN and RFID network integration. In *2015 2nd International Conference on electronics and communication systems (ICECS)*, pp. 497–502.

Novianti, D., Arkeman, Y., Almunawar, M. N., Haditjaroko, L., & Ismayana, A. (2020). Designing a transparent distributed systems for Halal supply chains using blockchain technology. *Journal of Business & Economic Analysis, 3*(2), 151–170.

Ometoruwa, T. (2018). *Solving the Blockchain trilemma: Decentralisation, security & scalability*. CoinBureau. Accessed May 22, 2019, from https://www.coinbureau.com/analysis/solving-blockchain-trilemma/

Opara, L. U. (2003). Traceability in agriculture and food supply chain: A review of basic concepts, technological implications, and future prospects. *Food, Agriculture & Environment, 1*(1), 101–106.

Park, K. (2018). Cryptocurrencies, blockchain is about to revolutionize the shipping industry. *Bloomberg*. Accessed July 22, 2019, from https://www.bloomberg.com/news/articles/2018-04-18/drowning-in-a-sea-of-paper-worlds-biggest-ships-seek-a-way-out

Pearson, S., May, D., Leontidis, G., Swainson, M., Brewer, S., Bidaut, L., Frey, J. G., Parr, G., Maull, R., & Zisman, A. (2019). Are distributed ledger technologies the panacea for food traceability? *Global Food Security., 20*, 145–149.

Perboli, G., Musso, S., & Rosano, M. (2018). Blockchain in logistics and supply chain: A lean approach for designing real-world use cases. *IEEE Access, 6*, 62018–62028.

Petersen, M., Hackuys, N., & See, V. B. (2018). Mapping the sea of opportunities: Blockchain in supply chain and logistics. *Information Technology, 60*, 1–8.

Pizzuti, T., Mirabelli, G., Sanz-Bobi, M. A., & Gomez-Gonzalez, F. (2014). Food track & trace ontology for helping the food traceability control. *Journal of Food Engineering, 120*(1), 17–30.

Pournader, M., Shi, Y., Seuring, S., & Koh, S. C. L. (2020). Blockchain applications in supply chains, transport and logistics: A systematic review of the literature. *International Journal of Production Research, 58*(7), 2063–2081.

Queiroz, M. M., Telles, R., & Bonilla, S. H. (2019). Blockchain and supply chain management integration: A systematic review of the literature. *Supply Chain Management, 25*(2), 241–254.

Rejeb, A. (2018). Blockchain potential in tilapia supply chain in Ghana. *Acta Technica Jaurinensis, 11*(2), 104–118.

Roger, E. M., Medina, U. E., Rivera, M. A., & Wiley, C. (2005). Complex adaptive systems and the diffusion of innovations. *The Innovation Journal: The Public Sector Innovation Journal, 10*(3), 1–26.

Rogers, E. M. (1995). *Diffusion of innovations* (4th ed.). The Free Press.

Rogerson, M., & Parry, G. C. (2020). Blockchain: Case studies in food supply chain visibility. *Supply Chain Management: An International Journal, 25*(5), 601–614.

Roth, A. V., Tsay, A. A., Pullman, M. E., & Gray, J. V. (2008). Unravelling the food supply chain: Strategic insights from China and the 2007 recalls. *Journal of Supply Chain Management, 44*(1), 22–39.

Ruiz-Garcia, L., Steinberger, G., & Rothmund, M. (2010). A model and prototype implementation for tracking and tracing agricultural batch products along the food chain. *Food Control, 21*(2), 112–121.

Russo-Spena, T., & Mele, C. (2012). "Five Co-s" in innovating: A practice-based view. *Journal of Service Management, 23*(4), 527–553.

Saberi, S., Kouhizadeh, M., Sarkis, J., & Shen, L. (2018). Blockchain technology and its relationships to sustainable supply chain management. *International Journal of Production Research, 57*(7), 2117–2135.

Salah, K., Nizamuddin, N., Jayaraman, R., & Omar, M. (2019). Blockchain-based soybean traceability in agricultural supply chain. *IEEE Access, 7*, 73295–73305.

Sander, F., Semeijn, J., & Mahr, D. (2018). The acceptance of blockchain technology in meat traceability and transparency. *British Food Journal, 120*(9), 2066–2079.

Saurabh, S., & Dey, K. (2021). Blockchain technology adoption, architecture, and sustainable agri-food supply chains. *Journal of Cleaner Production, 284*, 1–13.

Sawhney, M., Walcott, R., & Arronitz, I. (2006). The 12 different ways for companies to innovate. *MIT Sloan Management Review, 46*(3), 75–81.

Schumpeter, J. A. (1934). *The theory of economics development*. Oxford University Press.

Seebacher, S., & Schuritz, R. (2017). Blockchain technology as an enabler of service systems: A structured literature review. *International Conference on Exploring Services Science, 279*, 12–23.

Seuring, S., & Gold, S. (2012). Conducting content-analysis based literature reviews in supply chain management. *Supply Chain Management, 17*(5), 544–555.

Seuring, S., & Muller, M. (2008). From a literature review to a conceptual framework for sustainable supply chain management. *Journal of Cleaner Production., 16*(15), 1699–1710.

SGS. (2018). *Understanding transparency and traceability in the supply chain*. Accessed June 17, 2019, from https://www.sgs.com/en/news/2018/08/understanding-transparency-and-traceability-in-the-supply-chain

Shanahan, C., Kernan, B., Ayalew, G., McDonnell, K., Butler, F., & Ward, S. (2009). A framework for beef traceability from farm to slaughter using global standards: An Irish perspective. *Computers and Electronics in Agriculture, 66*(1), 62–69.

Shaw, G. (2018). 17 of the biggest food recalls of the year. *INSIDER*. Accessed February 07, 2019, from https://www.thisisinsider.com/food-recalls-2018-2018-12

Shaw, S., Grant, D. B., & Mangan, J. (2021). *A supply chain practice-based view of enablers, inhibitors and benefits for environmental supply chain performance measurement* (Vol. 32, pp. 382–396). University of Hull.

ShipChain. (2018). *Disrupting Transport & Logistics on the Blockchain*. ShipChain. Accessed October 30, 2018, from https://shipchain.io

Silva, M. E., Pereira, S. C. F., & Cold, S. (2018). The response of the Brazilian cashew nut supply chain to natural disasters: A practice-based view. *Journal of Cleaner Production, 204*, 660–671.

Skender, H. P., & Zaninovic, P. A. (2020). Perspectives of blockchain technology for sustainable supply chains. *Integration of Information Flow for Greening Supply Chain Management, 2020*, 77–92.

Song, L., Wang, X., & Merveille, N. (2020). Research on blockchain for sustainable E-Agriculture. In *2020 IEEE Technology & Engineering management conference (TEMSCON)*.

Spekman, R. E., Kamauff, J. W., Jr., & Myhr, N. (1998). An empirical investigation into supply chain management: A perspective on partnerships. *International Journal of Physical Distribution & Logistics Management, 28*(8), 630–650.

Stevens, G. C. (1989). Integrating the supply chain. *International Journal of Physical Distribution & Materials Management, 19*(8), 3–8.

Stranieri, S., Riccardi, F., Meuwissen, M. P. M., & Soregaroli, C. (2021). Exploring the impact of blockchain o the performance of agri-food supply chains. *Food Control, 119*, 1–12.

Su, S. I. I., Gammelgaard, B., & Yang, S. L. (2011). Logistics innovation process revisited: Insights from a hospital case study. *International Journal of Physical Distribution & Logistics Management, 41*(6), 577–600.

Sun, S., Wang, X., & Zhang, Y. (2017). Sustainable traceability in the food supply chain: The impact of consumer willingness to pay. *Sustainability, 9*(6), 999.

Tan, B., Yan, J., Chen, S., & Liu, X. (2018). The impact of blockchain on food supply chain: The case of Walmart. In *Proceedings of the international conference on smart Blockchain 2018, Tokyo, Japan*, pp. 166–177.

Tan, A., Gligor, D., & Ngah, A. (2020). Applying blockchain for Halal food traceability. *International Journal of Logistics: Research and Applications*, 1–19.

Thiruchelvam, V., Mughisha, A. S., Shahpasand, M., & Bamiah, M. (2018). Blockchain-based technology in the coffee supply chain trade: Case of Burundi coffee. *Journal of telecommunication, Electronic and Computer Engineering, 10*, 3–2.

Tian, F. (2016). An Agri-food supply chain traceability system for China based on FRID& Blockchain technology. In *2016 13th international conference on service systems and service management (ICSSSM)*, pp. 1–6.

Tian, F. (2017). A supply chain traceability system for food safety based on HACCP, Blockchain & Internet of Things. In *2017 International Conference on Service Systems and Service Management*, pp. 1–6.

Tieman, M., & Darun, M. R. (2017). Leveraging blockchain technology for halal supply chains. *Islam and Civilisational Renewal, 8*, 547–550.

Ting, S. L., Tse, Y. K., Ho, G. T. S., Chung, S. H., & Pang, G. (2014). Mining logistics data to assure the quality in a sustainable food supply chain: A case in the red wine industry. *International Journal of Production Economics, 152*, 200–209.

Treacy, R., Humphreys, P., Mcivor, R., & Lo, C. (2019). ISO14001 certification and operating performance: A practice-based view. *International Journal of Production Economics, 208*, 319–328.

Trienekens, J., & Zuurbier, P. (2008). Quality and safety standards in the food industry, developments and challenges. *International Journal of Production Economics, 113*(1), 107–122.

Trienekens, J., van Uffele, R., Debaire, J., & Omta, O. (2008). Assessment of innovation and performance om the fruit chain: The innovation-performance matrix. *British Food Journal, 110*(1), 98–127.

Tripoli, M., & Schmidhuber, J. (2018). *Emerging opportunities for the application of Blockchain in the agri-food industry*. Food and Agriculture Organization of the United Nations. Accessed December 09, 2019, from https://research4agrinnovation.org/wpcontent/uploads/2018/09/emerging_opportunities_for_the_application_of_blockchain_in_the_agri-food_industry_final_0.pdf

Tse, D., Zhang, B., Yang, Y., Cheng, C., & My, H. (2017). Blockchain application in food supply information security. In *Proceedings of the IEEE International Conference on Industrial Engineering and Engineering Management (IEEM 2017)*, pp. 1357–1361.

Tseng, J. H., Liao, Y. C., Chong, B., & Liao, S. W. (2018). Governance on the drug supply chain via Gcoin Blockchain. *International Journal of Environmental Research and Public Health, 15*(6), 1055.

Tsolakis, N., Niedenzu, D., Simonetto, M., Dora, M., & Kumar, M. (2020). Supply network design to address United Nations Sustainable Development Goals: A case study of blockchain implementation in Thai fish industry. *Journal of Business Research, 2020*, 1–25.

Tyco Integrated Security. (2012). Recall: The food industry's biggest threat to profitability. *Food Safety Magazine*. Accessed November 25, 2018, from https://www.foodsafetymagazine.com/signature-series/recall-the-food-industrys-biggest-threat-to-profitability/

United Nations Global Compact. (2014). *A guide to traceability*. Accessed February 07, 2019, from https://www.unglobalcompact.org/docs/issues_doc/supply_chain/Traceability/Guide_to_Traceability.pdf

Van Hilten, M., Ongena, G., & Ravesteijn, P. (2020). Blockchain for organic food traceability: Case studies on drivers and challenges. *Policy and Practice Review, 3*, 1–13.

Van Hoek, R. (2019). Exploring blockchain implementation in the supply chain. *International Journal of Operations & Production Management, 39*(6/7/8), 829–859.

Verhoeven, P., Sinn, F., & Herden, T. T. (2018). Examples from Blockchain implementations in logistics and supply chain management: Exploring the mindful use of a new technology. *Logistics, 2*(3), 1–19.

Wagner, S. M. (2008). Innovation management in the German transportation industry. *Journal of Business Logistics, 29*(2), 215–232.

Wang, Y., Han, J. H., & Davies, P. (2019a). Understanding blockchain technology for future supply chains: A systematic literature review and research agenda. *Supply Chain Management, 24*(1), 62–84.

Wang, Y., Singgih, M., Wang, J., & Rit, M. (2019b). Making sense of blockchain technology: How will it transform supply chains? *International Journal of Production Economics, 211*, 221–236.

WCED (World Commission on Environment and Development). (1987). *Report of the World Commission on Environment and Development: Our common future*. Accessed August 25, 2020, from https://sustainabledevelopment.un.org/content/documents/5987our-common-future.pdf

Weber, I., Gramoli, V., Ponomarev, A., Staples, M., Holz, R., Tran, A. B., & Rimba, P. (2017). On availability for Blockchain-based system. In *2017 IEEE 36th Symposium on Reliable Distributed System*, pp.64–73.

WHO. (2020). *Food safety*. Accessed August 25, 2019, from https://www.who.int/news-room/fact-sheets/detail/food-safety

Wilson, T. P., & Clark, W. R. (1998). Food safety and traceability in the agricultural supply chain: Using the internet to deliver traceability. *Supply Chain Management, 3*(3), 127–133.

Winter, S. G. (1964). Economic 'natural selection' and the theory of the firm. *Yale Economics Essays, 4*, 225–575.

Winter, S. G. (1994). Organizing for continuous improvement: Evolutionary theory meets the quality revolution. In B. JAC & J. V. Singh (Eds.), *Evolutionary dynamics of organizations* (pp. 90–108). Oxford University Press.

Wognum, P. M., Bremmers, H., Trienekens, J. H., & van der Vorst, J. G. A. J. (2011). Systems for sustainability and transparency of food supply chains – Current status and challenges. *Advanced Engineering Information, 25*(1), 65–76.

World Economic Forum (WEF). (2019). *Innovation with a purpose: Improving traceability in food value chains through technology innovations*. System Initiative on Shaping the Future of Food. Accessed April 17, 2019, from http://www3.weforum.org/docs/WEF_Traceability_in_food_value_chains_Digital.pdf

Wu, H., Li, Z., King, B., Miled, Z. B., Wassick, J., & Tazelaar, J. (2017). A distributed ledger for supply chain physical distribution visibility. *Information, 8*(4), 137.

Xu, L. D., Xu, E. L., & Li, L. (2018). Industry 4.0: State of the art and future trends. *International Journal of Production Research, 56*(8), 2941–2961.

Yakovleva, N. (2007). Measuring the sustainability of the food supply chain: A case study of the UK. *Journal of Environmental Policy and Planning, 9*(1), 75–100.

Yakovleva, N., & Flynn, A. (2004). Innovation and sustainability in the food system: A case of chicken production and consumption in the UK. *Journal of Environmental Policy & Planning, 6*(3–4), 227–250.

Yiannas, F. (2018). A new era of food transparency powered by blockchain. *Innovations: Technology, Governance, Globalization, 12*(1–2), 46–56.

Yu, M., & Nagurney, A. (2013). Competitive food supply chain networks with application to fresh produce. *European Journal of Operational Research, 224*(2), 273–282.

Zhao, J. L., Fan, S., & Yan, J. (2016). Overview of business innovations and research opportunities in blockchain and introduction to the special issue. *Financial Innovation, 2*(28), 1–7.

Zhao, G., Liu, S., Lopez, C., Lu, H., Elgueta, S., Chen, H., & Boshkoska, B. M. (2019). Blockchain technology in agri-food value chain management: A synthesis of applications, challenges and future research directions. *Computer in Industry, 109*, 83–99.

Chapter 3
Research Methodology

This chapter discusses the research strategies including research method, case selection, data collection, and data analysis. Ethical issues and quality criteria are also covered.

3.1 Research Method

This section discusses the detailed case study method, case design, data collection, data analysis, followed by the ethical issues and case study validation.

3.1.1 Qualitative Research

"Qualitative research is a form of social science where focus is on understanding people's world, interpreting their experiences and making sense out of it", according to Gupta and Awasthy (2015, p. 14). This means that qualitative study sets out to investigate, understand, and interpret certain phenomena in a natural setting (Denzin & Lincoln, 1994; Gupta & Awasthy, 2015). According to various researchers, qualitative research can provide deep and rich insights into the phenomenon, and benefits both practitioners and researchers (Boje, 2001; Van Maanen, 1979; Rowan, 2006). Qualitative research is effective in finding intangible influencing factors such as gender, cultural background, ethnicity, and religion. It tends to be used for theory development as it can provide details and description of the research context, human experience, beliefs, and emotions. Qualitative research is useful for various subjects such as information systems (Gupta & Awasthy, 2015).

There are five features of qualitative research which are summarised in Table 3.1 (Gupta & Awasthy, 2015; Maxwell, 2008): rich description, holistic perspective,

C. Zhang et al., *Blockchain Applications in Food Supply Chain Management*, Contributions to Management Science,
https://doi.org/10.1007/978-3-031-27054-3_3

Table 3.1 Five features of qualitative research (Source: Gupta & Awasthy, 2015, p. 16)

Rich description	Findings are descriptive, direct quotes are presented to capture the participants' experiences; focus is laid on words and pictures than on numbers.
Holistic perspective	This perspective seeks to understand the whole picture of the social context under investigation. It helps to understand complex interdependencies
Exploratory study	Qualitative research helps in developing hypothesis for further investigation, especially when a phenomenon is less researched and specific research questions and hypothesis are unclear. It attempts to answer what, how, and why questions to get a deeper, multifaceted understanding of the phenomena.
Understanding dynamism	Social relations and interactions are always in the state of confluence. It acknowledges the dynamism in interactions and social discourses.
Conceptual clarity/building theories	It helps to recognise the unexplored or unanticipated dimensions.

exploratory study, understanding dynamism, and conceptual clarity/building theories (Table 3.1). Qualitative research can provide a rich descriptive content based on the participants' quotations to understand the whole picture and social interactions involved and help to develop concepts. In some new research areas, qualitative research can also be used to develop hypotheses for further research.

3.1.2 Case Study

Case study method is applied in this research. According to Yin (2018, p. 15), a case study is

> an empirical inquiry that investigates a contemporary phenomenon in depth and within its real-life context, especially when the boundaries between phenomenon and context are not clearly evidenced.

In an early study, Benbasat (1987, p. 370) suggested that

> a case study examines a phenomenon in its natural setting, employing multiple methods of data collection to gather information from one or a few entities (people, groups, or organizations).

Voss et al. (2002, p. 197) asserted that "Case research is the method that uses cases studies as its basis". The case study method is employed in this paper to develop the understanding of the phenomenon under investigation and answer the "how" and "why" questions (McCutcheon & Meredith, 1993; Meredith, 1998; Saunders et al., 2015; Voss et al., 2002; Yin, 2018). The case study is usually deployed to develop new theories or test particular aspects of existing theories (Benbasat et al., 1987; Eisenhardt, 1989; Meredith, 1998; Richardt & Cook, 1979; Voss et al., 2002). Stranieri et al. (2021) suggested that in the early phase of blockchain technology, the explorative case study is an appropriate research method. The approach reflects on: (1) how blockchain applications can benefit food supply chain and improve food supply chain traceability, and (2) how companies deal with the challenges they encounter when engaging with blockchain applications.

There are many strengths in case-based research that were proposed by previous studies (Benbasat et al., 1987; Meredith, 1998; Stuart et al., 2002; Voss et al., 2002). First, it provides a more comprehensive way to understand certain theory from observing empirical practices. Stuart et al. (2002, p. 420) argued that case-based research "is a scientific approach that attempts to ground theoretical concepts with reality". Second, by being in the field and experiencing what is going on in the industry, it is possible to generate a new understanding of the research questions and gain more insights of the answers (Voss et al., 2002). Stuart et al. (2002) also confirmed that case studies can help to enhance the understanding of the phenomenon. Third, it is possible to reveal some unexpected and unknown influencing factors in the phenomena, which would lead to a better research design. Case-based research can deal with multiple variables and complex environment (Stuart et al., 2002). Last, the explanation of the phenomenon can be enriched. The case study can be used for theory and hypotheses testing (Meredith, 1998).

The complex system implementation process, unexpected events, multiple influencing factors, and complex operating environment all determine the suitability of using case studies in information system research (Benbasat et al., 1987; Walsham, 1995). For certain topics that have theories in their early stage, and practical experiences are important, case research is particularly useful (Benbasat et al., 1987). Benbasat et al. (1987) suggested that some case study topic examples include the impact of the information system and the influencing factors. In blockchain studies, the case study can not only provide an opportunity to have a better understanding of the research but also provide deeper insights into the research questions.

Compared to single-case design, multiple-case design has distinct advantages, and this is adopted in this research. "When multiple settings are investigated to help extend the generalisability of the results, we call this a multiple case study" (Meredith, 1998, p. 443). Multiple cases can provide more compelling evidence with more robust findings (Herriott & Firestone, 1983). Benbasat et al. (1987) and Meredith (1998) suggested that the multiple case study can extend the generalisability of the single case study. Benbasat et al. (1987) also suggested that multiple case studies are more suitable for research that intends to build or test theories. A key advantage is to allow deep insights into the phenomenon under study. In addition, the multiple case study provides opportunities to conduct data collection from different sources of evidence to develop converging lines of inquiry (Yin, 2018). It is particularly suitable for the topic when there is some knowledge and understanding, but most of it still waits to be explored (Meredith, 1998).

3.2 Case Selection

This research uses both convenient and theoretical sampling approaches to investigate how food companies adopt the blockchain applications in supply chain management. According to Corbin and Strauss (2008, p. 143), theoretical sampling is "a method of data collection based on concepts/themes derived from data". It is particularly useful for relatively new research fields that do not attract many existing theories. It allows more discoveries with responsiveness and flexibility. The potential candidates of this research are the food companies that are using blockchain applications and playing a leading role in their food supply chain.

The case selection process (Fig. 3.1) has five steps. First, a sample pool was identified by reading relevant news, reports, and academic papers. A few companies that had launched blockchain projects in food supply chain were identified, such as IBM food trust, Walmart, and Provenance.

Next, the cover letter (Appendix 1) and interview protocol (Appendix 2) were sent to the potential companies via various channels, for example by email, social media, LinkedIn, and online query. The cover letter was developed to give the potential candidates the basic information of this research, such as research

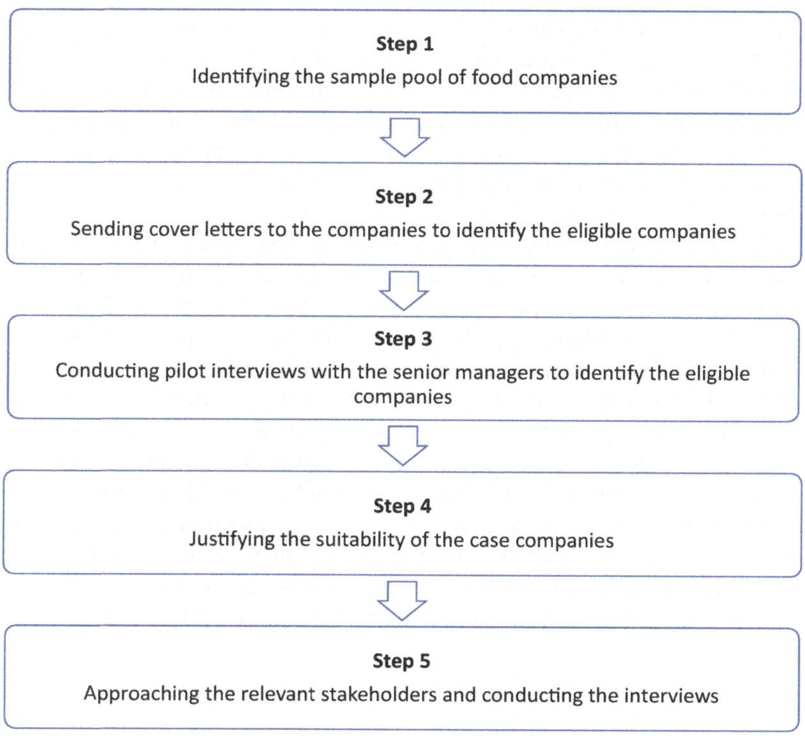

Fig. 3.1 Basic information of case companies

Table 3.2 General information of three selected companies (Data as of June 2021)

General information	BeefLedger	W company China branch	FairChain
Company location	Australia	China	Netherland
Product and supply chain characters	Cross-border beef supply chain	Food retailing	Outsourced coffee supply chain
Company demand	Food traceability and Supply chain transparency	Food traceability and supplier management	Value distribution & supply chain sustainability
Target market	Chinese consumers	Chinese consumers	Netherland coffee drinkers
Project start time	2017	2018	2018
Project stage	Mature	Phase 2	Mature
Number of employees	12	96,600	6
Website	https://BeefLedger.io	N/A	https://FairChain.org

background, author's background, contact details, and sample interview questions. The interview protocol includes blockchain, food supply chain, and innovation-related questions, and it targets focal companies and tech support companies. From the cover letter, the companies could decide if they wanted to participate in this research. At this step, more than 30 companies were contacted.

When companies showed strong interest in this research and agreed to talk further about their blockchain project, a pilot interview was held with their senior managers to identify more details of their projects. In addition, some industry experts were also contacted and interviewed to gain some views from different angles based on their industry experience of various blockchain projects. For example, an interview was held with a technology expert who suggested that blockchain is better used for value distribution purposes. Moreover, a blockchain expert who works in the shipping industry believes that blockchain is a brilliant solution for food supply chains.

The eligible companies should: (1) show strong interest to participate in this research; (2) have an ongoing blockchain project and have relevant experience in food blockchain projects; (3) have more than one participant that is willing to take part in this research; (4) have engaged multiple supply chain partners; and (5) be willing to share their internal documents. After the pilot interviews, three eligible companies were identified for the case studies in this research. BeefLedger is conducting a blockchain project in a cross-border beef supply chain, W company is a food retailer focusing on adopting blockchain for food traceability and supplier management, and FairChain is focusing on applying blockchain to achieve value redistribution along a more sustainable coffee supply chain. The basic information of the companies is listed in Table 3.2, including their targets, supply chain categories, number of employees, and websites. The last step of the case study is officially conducting interviews with relevant supply chain stakeholders.

3.3 Data Collection

This research uses primary data as the main data source by interviewing the relevant practitioners. As mentioned before, blockchain is in its infancy, so there is limited research and information on cases that use it. Thus, most existing research papers provided conceptual frameworks based on theories. In this case, conducting interviews with experts and case companies can efficiently help to gain deep insights of the research topic. Yin (2018) proposed six data sources for case studies—documents, archival records, interviews, direct observation, participant observation, and physical artefacts. Among the six sources, Walsham (1995) argued that the interview is the primary data source for case studies, as it can provide broad access to the participants' perceptions and thoughts. To learn the details of blockchain implementation process and blockchain impacts on food supply chain, the desired data collection method is to directly interview the project participants.

Semi-structured interviews are adopted as the primary data sources in this research. Semi-structured interviews can provide access to the background, and investigate the relationships between variables (Saunders et al., 2015). According to Saunders et al. (2015), semi-structured interviews usually have key themes and cover key questions. Some extra questions can be added along the interview to enhance the understanding and to extend the research discussion. Compared to a standardised structured interview, the semi-structured interview has more flexibility, and the questions can be varied depending on the interviewees (Pagell & Wu, 2009). Different companies usually have different capabilities; therefore, semi-structured interviews allow the interviewer to focus on the uniqueness of each company. Additional questions can also be added where needed. This also means that there are chances to explore some new areas during the discussion (McCutcheon & Meredith, 1993). To study blockchain, semi-structured interviews not only provide a better understanding of the research context, such as the company background and experience with blockchain implementation but also gives deeper insights of the research questions such as blockchain adoption influencing factors and some unexpected variables.

The interview protocol has sample questions for selected companies and their stakeholders. The questions relate to food supply chain, blockchain, and innovation to answer the research questions. For example, questions about the supply chain issues that the company is facing and the potential reasons that caused the issues can provide a view to the reasons of why company decides to adopt blockchain, will answer the first research question (RQ1) (Why do companies apply blockchain in food supply chain management and based on what threats or opportunities in the food supply chain?). The questions around blockchain provide details to the blockchain implementation process, such as issues and challenges during implementation. These questions generate answers to RQ1. There are also some questions about other innovations, such as what innovations can replace or complement blockchain. The questions about innovation in general can help to explore the potentials and values of blockchain.

In total 23 interviews were conducted (Table 3.3). Twelve out of 23 interviews were with the participants from BeefLedger, W company, and FairChain. Eleven interviews were held with industry experts from various backgrounds which triangulated the data source and validated the research findings. Due to the COVID-19 restrictions (from 2020 to present), all the interviews were conducted by telephone, or via online meeting software such as Microsoft Teams and Zoom. The basic information of interviewees including their job roles, dates of the interview, and length of the interviews is listed in Table 3.3. The interview languages are English and Chinese. All interviews were recorded and transcribed for data analysis purpose. The English and Chinese audios were transcribed by Rev.com (https://www.rev.com/transcription) and by Iflyrec (https://www.iflyrec.com) respectively. During the interviews, a few internal documents were collected from the case companies (Table 3.4). Moreover, the researcher also participated in several online conference meetings besides the formal interviews such as "Webinar: blockchain- the China Way" and "What you might not know about PoA/Blockchain consensus algorithms: a brief intro to PoA2.0".

Saunders et al. (2015) suggested that secondary data are often used in case study or survey research and can provide extra discoveries. Thus, in this research, information from reports, news, and websites was collected to complement the data resources. For example, the websites of the case companies contain rich information including company background and blockchain project reports.

3.4 Data Analysis Strategy

Data were coded soon after each interview to lay a foundation for further data collection (Corbin & Strauss, 2008). Open coding was first applied in followed by axial coding. Open coding is used for "breaking data part and delineating concepts to stand for blocks of raw data" (Corbin & Strauss, 2008, p. 195). In other words, open coding is used for conceptualising data. It is a process that allows theories to emerge from the raw data (Khandkar, 2009; Moghaddam, 2006). From the literature review, a few constructs were identified. For instance, the construct of the innovation process is: (1) setting the stage activities, (2) customer clue-gathering activities, (3) negotiating, clarifying, and reflecting activities, and (4) inter-organisational learning (Flint et al., 2005). The barriers during blockchain implementation are intra-organisational barriers, inter-organisational barriers, systems-related barriers, and external barriers (Saberi et al., 2018). The influencing factors are capabilities, collaboration, technology readiness, and external environment (Hastig & Sodhi, 2020). Table 3.5 provides some of the coding examples. Detailed data analysis and comparison are presented in subsequent chapters.

Axial coding is applied to reduce the number of data labels after open coding. It is used to explore data in depth and reassemble data. Axial coding can find out the relationships between variables and concepts, which lays a foundation for further extracting theories. During data coding, the coding should be carried on

Table 3.3 List of all interviews

Company	Chain level	Title/name	Language	Date	Interview length (h)
BeefLedger	Focal company/ Blockchain technology provider	Chairman—Warwick	English	08.04.2020	1:08
	Focal company/ Blockchain technology provider	Chairman—Warwick	English	04.05.2020	1:19
	Focal company/ Blockchain technology provider	Smart contracts & supply chain risk researcher—Dr. Cao	Chinese	28.05.2020	1:08
	Focal company/ Blockchain technology provider	Smart contracts & supply chain risk researcher—Dr. Cao	Chinese	05.06.2020	1:55
W company	Blockchain technology provider	Co-Founder Chief Financial Officer—Mr. Zhang	Chinese	26.05.2020	1:59
	Blockchain technology provider	Project manager—Mr. Cheng	Chinese	23.06.2020	1:33
	Blockchain technology provider	Project manager—Mr. Cheng	Chinese	18.07.2020	1:37
	Consulting	Senior Manager—Mr. Chen	Chinese	21.07.2020	1:10
	Blockchain technology provider	Co-Founder Chief Financial Officer—Mr. Zhang	Chinese	07.05.2021	1:47
FairChain	Focal company/ Blockchain technology provider	Platform architect—Mr. A	English	29.11.2019	00:58
	Focal company/ Blockchain technology provider	Chief technology officer—MR. R	English	02.11.2020	1:05
	Leeds University	Academic advisor—Deepak	English	03.04.2020	00:42
Lloyd's registers	Blockchain technology provider for shipping industry	Innovation Owner Marine & Offshore—Gary	English	11.12.2019	00:54
	Blockchain technology provider for shipping industry	Vice President for Food, Beverage & Sustainability—Vincent	English	04.03.2020	00:36
VastChain	Blockchain technology provider	Mr. Chen	Chinese	23.12.2019	00:41
Blockchain expert	Blockchain technology provider	Mr. Fan	Chinese	05.01.2020	00:35

(continued)

Table 3.3 (continued)

Company	Chain level	Title/name	Language	Date	Interview length (h)
SIM	Blockchain technology provider for food supply chain	CTO-Caspar	English	20.03.2020	1:13
FinTech4Good	Blockchain technology provider	President—Mr. Zhang	Chinese	16.06.2020	1:08
Blockchain expert	Blockchain technology provider	Mr. Luo	Chinese	07.07.2020	00:46
AgLive	Beef supply chain	Executive Director—Mark TooHey	English	05.08.2020	00:49
Escavox	IoT provider	Chief Executive officer—Luke Wood	English	12.08.2020	1:16
Rolar de cuyo	Olive oil supply chain	Director-Guillermo	English	20.11.2020	1:16
Two Hands	Lobster supply chain	Business leader—Stuart	English	25.11.2020	1:40

Table 3.4 Additional sources

Companies	Additional sources
VeChain	Conference meeting: Webinar: Blockchain—the China way
	Conference meeting: What you might not know about PoA/Blockchain consensus algorithms: a brief intro to PoA2.0
	Company website
	Internal reports and documents
BeefLedger	Internal reports and documents
	Company website
FairChain	Internal reports and documents
	Company website

continuously until saturation is reached. This was applied to the project outcome construct. For example, are six blockchain adoption benefits (efficiency, information transparency and supply chain visibility, information authenticity and accountability, digitisation, resilience, sustainability) were obtained from various literature and interview data.

There are within-case analysis and cross-case analysis in the case analysis. Data collected from each case were first subjected to within-case analysis, followed by cross-case analysis. Within-case analysis refers to the summary of the key data in each case in order to have a better understanding and general insights of each case (Eisenhardt, 1989). Chapters 4, 5, and 6 present the within-case analysis of this research. The within-case analysis follows the same structure by introducing the project background and supply chain issue, following by the different blockchain

Table 3.5 Coding scheme of data analysis

Constructs	Sub constructs	Operationalised constructs and definitions	Example quotations	References
Innovation process	Setting the stage activities	Supply chain issues: the issues that focal companies were facing and hope to address by blockchain, including supply chain unsustainability, supply chain inefficiency, and food security issue. Key business partners: gathering key business partners before taking a step further.	"The product origins used to be manually entered, and can be changed in the background anytime. This kind of data changing is happening quite often in the food industry. For example, we all know that pears from Xinjiang are delicious, so the suppliers can change the pear origins to Xinjiang. This is the weakness of the traditional traceability method".—Mr. Feng, senior manager at PwC food & beverage retail. "I source it via a company called KripC, and they have extensive knowledge on Hyperledger and how to build blockchains on that. And together we design blockchain. So, we do the service design together and they build the intellectual properties for FairChain".—Mr. R, the chief technology officer of FairChain Foundation	Flint et al. (2005), Su et al. (2011), Bjorklund and Forslund (2018)
	Customer clue-gathering activities	Data collection: The methods the focal companies use to collect data and	"We did about a year of internal research and development	

(continued)

Table 3.5 (continued)

Constructs	Sub constructs	Operationalised constructs and definitions	Example quotations	References
		conduct research before implementing blockchain, such as conducting research and running pilots.	before launching the project. So we officially launched the project around October, November, 2017. So we began our internal research and development and business case mapping in 2016".—Warwick Powell, Chairman of BeefLedger "The first mile means that the farmer has a face. It's not just a farmer, but we know all our farmers by name, we have their faces and their profiles. We know what they're producing, how big their families are. And now we're capturing data that will contribute to our long-term ambition to bring them to a living income level".—Mr. R, the chief technology officer of FairChain Foundation	
	Negotiating, clarifying, and reflecting activities	Stakeholder engagement: the behaviours the focal companies engaged in to persuade and attract stakeholders to use blockchain, such as supplier education.	"With such an 'trust-free' area, suddenly, consumers bought out all of these products. Consumers think that scanning and seeing information is a fun thing to do, and	

(continued)

Table 3.5 (continued)

Constructs	Sub constructs	Operationalised constructs and definitions	Example quotations	References
			they believe what W company suggests. With these special arrangements, more suppliers are willing to join in. It starts to become a positive cycle and brings in more suppliers".—Mr. Zhang, Co-founder of VeChain "Suppliers will not give sensitive information to W company. Only the information for quality control, and information they are willing to show consumers are on the chain. For example, inspection reports, annual audit reports and qualification certificates, this non-sensitive information".—Mr. Cheng, the project manager in VeChain "We work on what we call a brand loyalty system. The idea behind this blockchain solutions is that we try to connect with consumers in a different way. And we're running experiments to see if providing insight of the value chain and value distribution can do	

(continued)

Table 3.5 (continued)

Constructs	Sub constructs	Operationalised constructs and definitions	Example quotations	References
			anything good for customer loyalty".—Mr. R, the chief technology officer of FairChain Foundation	
	Inter-organisational learning	Future: the plans that focal companies hope to implement in the future, such as using smart contract, and engaging with wider stakeholders via in-depth inter-organisational learning.	"For example, in the past, it usually has to conduct a multi-party verification to confirm the receiving of goods, and then payment can be initiated. The whole process is long. But if they start to use smart contract, as soon as suppliers ship the products and W company's distribution centres receive the products with the right amount and appropriate condition, the payment can be initiated. This is because the data of the products are recorded on the blockchain platform permanently. So, people have trust and no need to confirm so many things again. In other words, once the smart contract is set up, we can synchronise the completion of the delivery and billing in both the financial and legal sense".—Mr.	

(continued)

Table 3.5 (continued)

Constructs	Sub constructs	Operationalised constructs and definitions	Example quotations	References
			Feng, senior manager at PwC food & beverage retail "We will discuss with W company gradually about what to do next. The only certain thing in the future is that W company has plans to involve more suppliers and products. The other things are constantly adjusting to achieve continuous improvement".—Mr. Zhang, Co-founder of VeChain	
Contingency variables	Critical influencing factors	Capabilities: a company's resources and knowledge to carry out the blockchain project. Collaboration: The leadership of the lead companies and acceptance of supply chain stakeholders. Technology readiness: the technological maturity to adopt and compatibility to the current system. External environment: the attitude and of third parties, and the market reaction to blockchain.	"Let's say we use our beef supply chain as an example to let others find blockchain is indeed useful. They can get more consumer information and understand the consumer's behaviour to promote their products. We hope they can see the value and will join us gradually".—Dr. Cao, BeefLedger researcher "Government actually pays attention to it. At the national level, companies are encouraged to do such things (blockchain-enabled traceability), because it is	Grawe (2009), Post et al. (2018), Hastig and Sodhi (2020), Yadav and Singh (2020)

(continued)

Table 3.5 (continued)

Constructs	Sub constructs	Operationalised constructs and definitions	Example quotations	References
			meaningful for people's livelihood".—Mr. Cheng, the project manager in VeChain "Because W company wants to make a faster move, they decided to cover all the costs, including supplier training and consulting costs".—Mr. Feng, senior manager at PwC food & beverage retail	
	Barriers	Intra-organisational barriers: the barriers happen within the focal companies. Inter-organisational barriers: The barriers happen within a supply chain among stakeholders. Systems related barriers: the barriers relate to technology. External barriers: the barriers relate to external parties such as government.	"The first category of challenges relates to the technology itself. The technology itself is a young technology. So there are always issues in terms of the building of complex systems with young technologies. The second issue relates to the availability of skilled people. So, to be able to develop the coding, the people need very specific skills and not just skilled in coding, but also a knowledge of other industries that is a significant challenge and something that affects the entire industry".—Warwick Powell, Chairman of BeefLedger "I have identified some potential	Saberi et al. (2018)

(continued)

Table 3.5 (continued)

Constructs	Sub constructs	Operationalised constructs and definitions	Example quotations	References
			challenges from FairChain case, which is that people who are using blockchain technology doesn't have a knowledge about it. So, lack of understanding of the technology is one of the major challenges particularly at the lower tier level".—Dr. Deepak, a researcher at the University of Leeds	
Supply Chain performance	Project outcomes	Efficiency Information transparency and supply chain visibility Information authenticity and accountability Digitisation Resilience Sustainability	"It can form interaction. For example, consumers can know the production process, and the producer can receive consumer feedback".—Dr. Cao, BeefLedger researcher "Supply chains are actually changed from chains to cycles; they are circuits, the economic circuits. So, every stage in the circuit has its own need for information and players who are engaged in each of those stages will perform different roles. The farmer can be validated in the network for information that is collected at a different point in the process. Nothing stopped somebody	Bromiley and Rau (2014), Carter et al. (2017), Bromer et al. (2019), Silva et al. (2018)

(continued)

Table 3.5 (continued)

Constructs	Sub constructs	Operationalised constructs and definitions	Example quotations	References
			participating as a validator in the network. Nothing stops somebody running a storage node to store a copy of the information for the whole network".—Warwick Powell, Chairman of BeefLedger "This is what I always say that what blockchain can do is bring two different worlds together. Traditionally, we have the world of money and we have the world of anti-counterfeiting. The money world, just money. The anti-counterfeiting world is all about regulations and policing. By using economics to change the incentive mechanisms and the transparency of information, we can actually drive behaviour more and more towards the legitimate trade and weed out the illegitimate trade".—Warwick Powell, Chairman of BeefLedger "That's why we call it radical transparency. The consumers will know in the value chain	

(continued)

Table 3.5 (continued)

Constructs	Sub constructs	Operationalised constructs and definitions	Example quotations	References
			about who owns what. We've got the supply chain solution, which basically provides provenance of the product. And it also makes you follow the money. And you can look back from your product that you bought on who did earn what".—Mr. R, the chief technology officer of FairChain Foundation	

project stages, the benefits that have been brought to the case companies, and the barriers encountered during implementation.

The cross-case analysis is to identify the commodities and differences of phenomenon, events, and processes. According to Yin (2018, p. 196), "the goal is to retain the integrity of the entire case and then to compare or synthesize any within-case patterns across the cases". Miles et al. (2013, p. 101) also suggested that the proposed of cross-case analysis is to find out the "generalisability or transferability to other contexts" and to deepen "the understanding and explanation". The cross-case analysis aims to aggregate and compare the findings across the cases. Chapter 7 presents the cross-case analysis and discussion by comparing the findings of the three cases with existing literature.

Table 3.6 lists the techniques for case analysis and how the techniques are used in this research. The techniques include list, category, compare and explain the interviews. For example, for chronologies, the interview list is drawn up to present when, and interview with who. Coding includes a coding list to sort out relevant data. In the case study research with more than one case, within-case and cross-case analyses are applied to compare and explain the cases.

3.5 Ethical Issues

When doing research, there are some general ethical requirements (Boeije, 2010): informed consent, privacy, confidentiality, and anonymity. *Informed consent* is used to ensure that research participants have full knowledge of the research purpose and

Table 3.6 Case study analysis techniques (Source: Ghauri, 2004)

Techniques for case analysis	Explanation	Representation
Chronologies	Narratives of the events that took place, organised by date	Interview list
Coding	Sorting data according to concepts and themes	Coding list
Clustering	Categorising cases according to common characteristics	Cluster contextual variables
Matrices	Explaining the interrelationship between identified factors	Within- and cross-case analyses
Pattern matching	Comparison between a predicted and an empirically based pattern	Within- and cross-case analyses

willing to participate. *Privacy* means that data should not be disclosed to others without agreement. *Confidentiality* refers to how the data are handled to protect privacy. *Anonymity* suggests that the private information of participants should be kept private with nothing that can identify the participant. In this case, an Ergo submission was approved for data collection by the Faculty of Social Science, University of Southampton. Informed consent forms were sent out to the companies to explain the research purposes before starting data collection. The use of data and the privacy of participants strictly followed the ethical guidelines.

3.6 Quality Assurance

McCutcheon and Meredith (1993) suggested that the contribution of case study research is largely determined by the quality of the research design and data analysis. The research design quality is determined by the research validity and reliability. Yin (2018) suggested four tests for case studies: construct validity, internal validity, external validity, and reliability. *Construct validity* refers to have the correct operational measures for the concepts being studied (Voss et al., 2002). To enhance the construct validity, the research can adopt more than one evidence source and let the informants review the draft to validate the information (Stuart et al., 2002). After data collection, the recording transcripts should be sent to the participants to ensure the accuracy of the information.

Internal validity aims to find cause-and-effect relationships between treatment and results (McCutcheon & Meredith, 1993). It largely depends on how rigorously the research has been performed, which requires extra attention during the research procedures. In most cases, using multiple methods and resources can greatly improve internal validity. In some complex situations, "pattern matching" can be one of the approaches for internal validity (Campbell, 1975; McCutcheon and Meredith, 1993). In other words,

> pattern matching involves the comparison of a predicted theoretical pattern with an observed empirical pattern. The underlying assumption is that human beings make sense of the world by comparing what they observe externally to internal mental models (Sinkovics, 2018, p. 468).

External validity relates to whether the research findings are generalisable. According to McCutcheon and Meredith (1993, p. 246) external validity is

> the extent to which findings drawn from studying one group are applicable to other groups or settings, is a special concern for case studies. Results can be tested and extended by replication, that is, through the investigation of other cases where results should be comparable.

The two parts of the replication procedure are literal replication and theoretical replication (Yin, 2018). *Literal* replication refers to when the case selections and predicted results are similar. *Theoretical* replication refers to when the cases are selected based on the different results' assumptions.

Reliability refers to whether the same finding can be found after demonstrating the same research operational procedures. In this case, sufficient explanations of the research process including research design and data collection should be included to show the dependability of the research findings (Benbasat et al., 1987; Saunders et al., 2015). A few methods can be used to improve information reliability (McCutcheon and Meredith, 1993; Stuart et al., 2002; Voss et al., 2002): using protocol, using various data-gathering methods, using digital recorder, having more than one researcher present, and using multiple measures.

In this research, the interview protocol questions (Appendix 2) were used as research guidance and a major way of enhancing the research reliability. The research database (recording, transcripts, reports, notes) was adequately developed as solid evidence and allows readers to follow and trace the case study procedures. Information was gathered from people in different job roles from different supply chain stakeholders to triangulate data and avoid bias. Additional interviews were conducted with industry experts to provide additional sources of information to triangulate the research findings. The applications of reliability and validity in this study are shown in Table 3.7.

3.7 Summary

This chapter presents the research method. A qualitative research method—multiple case study method is going to be applied to explore blockchain implementations in food supply chain industry. This chapter presents the case selection and data collection process in detail. Moreover, data analysis and coding schemes is also discussed. There are also discussions about ethical issues and quality assurance.

Table 3.7 Reliability and validity in case research (Source: Yin, 2018)

Tests	Application in this study
Construct validity	1. Multiple sources of evidence including semi-structured interviews, secondary data 2. A chain of evidence: multiple informants in focal companies, and multiple informants at suppliers/non-traditional supply chain partners 3. Review of findings by uninvolved senior academics
Internal validity	1. Structured data coding and analysis 2. Development of propositions based on a chain of evidence
External validity	1. Theoretical sampling approach 2. Thick descriptive data 3. Participate in companies' online meeting/sessions
Reliability	1. Use case study protocol to guide research and analysis 2. Develop case study database including recordings, transcripts, internal documents, academic case studies, news coverage 3. Iterative discussions with uninvolved senior academics 4. Feedback from senior interviewees

References

Benbasat, T., Goldstein, D. K., & Mead, M. (1987). The case research strategy in studies of information system. *MIS Quarterly., 11*(3), 369–386.

Bjorklund, M., & Forslund, H. (2018). Exploring the sustainable logistics innovation process. *Industrial Management & Data Systems, 118*(1), 204–217.

Boeije, H. (2010). *Analysis in qualitative research*. Sage.

Boje, D. M. (2001). *Narrative methods for organizational & communication research*. Sage.

Bromer, J., Brandenburg, M., & Gold, S. (2019). Transforming chemical supply chains toward sustainability – A practice-based view. *Journal of Cleaner Production, 236*, 1–16.

Bromiley, P., & Rau, D. (2014). Research prospective: Towards a practice-based view of strategy. *Strategic Management Journal, 35*, 8.

Campbell, D. T. (1975). Degrees of freedom and the case study. *Comparative Political Studies, 8*(3), 178–193.

Carter, C. R., Kosmol, T., & Kaufmann, L. (2017). Toward a supply chain practice view. *Journal of Supply Chain Management, 53*(1), 114–122.

Corbin, J., & Strauss, A. (2008). *Basic of qualitative research: Grounded theory procedures and techniques* (3rd ed.). Sage.

Denzin, N. K., & Lincoln, Y. S. (1994). *Handbook of qualitative research*. Sage.

Eisenhardt, K. M. (1989). Building theory from case study research. *Academy of Management Review, 14*(4), 532–550.

Flint, D. J., Larsson, E., Gammelgaard, B., & Mentzer, J. T. (2005). Logistics innovation: A customer value-oriented social process. *Journal of Business Logistics, 26*(1), 113–147.

Ghauri, P. (2004). Designing and conducting case studies in international business research. In R. Marschan-Piekkari & C. Welch (Eds.), *Handbook of qualitative research methods for international business* (pp. 109–124). Edward Elgar.

Grawe, S. J. (2009). Logistics innovation: A literature-based conceptual framework. *The International Journal of Logistics Management, 20*(3), 360–377.

Gupta, R. K., & Awasthy, R. (2015). *Qualitative research in management: Methods and experiences*. Sage.

Hastig, G. M., & Sodhi, M. S. (2020). Blockchain for supply chain traceability: Business requirements and critical success factors. *Production and Operations Management, 29*(4), 935–953.

Herriott, R. E., & Firestone, W. A. (1983). Multisite qualitative policy research: Optimizing description and generalizability. *Educational Researcher, 12*(2), 14–19.

Khandkar, S. H. (2009). *Open coding*. Accessed August 12, 2019, from http://pages.cpsc.ucalgary.ca/~saul/wiki/uploads/CPSC681/opencoding.pdf

Maxwell, I. A. (2008). *Qualitative research design*. Sage.

McCutcheon, D. M., & Meredith, J. R. (1993). Conducting case study research in operations management. *Journal of Operations Management, 11*(3), 239–256.

Meredith, J. (1998). *Building operations management theory through case and field research* (Vol. 16, pp. 441–454). Wiley.

Miles, M. B., Huberman, A. M., & Saldana, J. (2013). *Qualitative data analysis: A methods sourcebook* (3rd ed.). Sage.

Moghaddam, A. (2006). Coding issues in grounded theory. *Issues in Educational Research, 16*(1), 47–58.

Pagell, M., & Wu, Z. (2009). Building a more complete theory of sustainable supply chain management using case studies of 10 exemplars. *Journal of Supply Chain Management, 45*(2), 37–56.

Post, R., Smit, K., & Zoet, M. (2018). *Identifying factors affecting blockchain technology diffusion*. Twenty-fourth Americas conference on information systems, New Orleans, pp. 1–10

Richardt, C. S., & Cook, T. D. (1979). Beyond qualitative vs quantitative methods. In C. S. Richardt & T. D. Cook (Eds.), *Qualitative and quantitative methods in evaluation research* (pp. 7–32). Sage.

Rowan, J. (2006). *Handbook of action research: Concise paperback edition* (pp. 106–116). Sage.

Saberi, S., Kouhizadeh, M., Sarkis, J., & Shen, L. (2018). Blockchain technology and its relationships to sustainable supply chain management. *International Journal of Production Research, 57*(7), 2117–2135.

Saunders, M., Lewis, P., & Thornhill, A. (2015). *Research methods for business students* (7th ed.). Harlow.

Silva, M. E., Pereira, S. C. F., & Cold, S. (2018). The response of the Brazilian cashew nut supply chain to natural disasters: A practice-based view. *Journal of Cleaner Production, 204*, 660–671.

Sinkovics, N. (2018). Pattern matching in qualitative analysis. In C. Cassell, A. L. Cunliffe, & G. Grandy (Eds.), *The Sage handbook of qualitative business and management research methods* (pp. 468–485). Sage.

Stranieri, S., Riccardi, F., Meuwissen, M. P. M., & Soregaroli, C. (2021). Exploring the impact of blockchain o the performance of agri-food supply chains. *Food Control, 119*, 1–12.

Stuart, I., McCutcheon, D., Handfield, R., McLachlin, R., & Samson, D. (2002). Effective case research in operations management: A process perspective. *Journal of Operations Management, 20*(5), 419–433.

Su, S. I. I., Gammelgaard, B., & Yang, S. L. (2011). Logistics innovation process revisited: Insights from a hospital case study. *International Journal of Physical Distribution & Logistics Management, 41*(6), 577–600.

Van Maanen, J. (1979). *Qualitative methodology*. Sage.

Voss, C., Tsikriktsis, N., & Frohlich, M. (2002). Case research. *Case Research in Operations Management, 22*(2), 195–219.

Walsham, G. (1995). Interpretive case studies in IS research: Nature and method. *European Journal of Information System, 4*(2), 74–81.

Yadav, S., & Sign, S. (2020). Blockchain critical success factors for sustainable supply chain. *Resources, Conservation & Recycling, 152*, 1–11.

Yin, R. K. (2018). *Case study research: Design and methods*. Sage.

Chapter 4
Case Analysis: BeefLedger

This chapter provides the findings of the case analysis of BeefLedger. It introduces the background of the company, the issues that the company wishes to address, different stages of blockchain implementation, the benefits of blockchain, and how the company overcame the barriers.

4.1 Background Information

This section presents the background information of the company and beef industry, issues in cross-border beef supply chain, and consumer concerns.

4.1.1 Company Background

BeefLedger is a start-up technology company formed by Sister City Partners Limited (a not-for-profit enterprise focused on regional economic development through enhanced cross-border trade and investment), with the support of a number of early-days investors. To encourage collaboration with research institutes, BeefLedger became a founding industry partner in the Queensland University of Technology-based Food Agility Cooperative Research Centre (CRC) project in 2017. Chairman Warwick Powell has worked in the beef industry for a long time and believes in blockchain's potential. BeefLedger aims to provide a platform to secure and streamline transactions between businesses involved in the production, processing, and distribution of livestock and beef products. This is based on the underlying protocol infrastructure—*Ethereum*, on which BeefLedger has deployed a layer-two application solution—the *Proof of Authority* (POA) ecosystem, which allows supply chain stakeholders to engage with the system. Due to their working

C. Zhang et al., *Blockchain Applications in Food Supply Chain Management*,
Contributions to Management Science,
https://doi.org/10.1007/978-3-031-27054-3_4

experience in the beef industry and in parallel with the financial services sector, Warwick Powell and BeefLedger director Charles Morris have also acquired sufficient resources to launch this project. BeefLedger (or affiliated entities) has its own agricultural assets; it purchases, processes, and sells cattle and works directly with partners in all aspects of the supply chain. In the beef supply chain, BeefLedger is principally a technology provider, but also plays the roles of producer, processor, and distributor to support first-hand research, development, and commercialisation.

There are 12 personnel in four departments: the technology development department (responsible for supply chain design, programming, blockchain scalability); the community value department (to help communities participate in the network and to use blockchain technology); the economic research department (using economics as incentives to engage more stakeholders), and the legal research department (focuses on policy and legal issues including data privacy). In addition, BeefLedger also works with a Beijing-based e-commerce ecosystem developer LibertyPost and Guangzhou-, Wuhu-, and Shenzhen-based local distributors to sell products and receive consumer feedback. BeefLedger also has a joint venture in South Africa. To date, BeefLedger's main consumers are middle-class Chinese consumers, who have more concerns over food safety and quality. For Chinese consumers, the two most important pieces of information on beef products are originality and transportation conditions, such as the temperature.

4.1.2 *Beef Industry in Australia*

The consumption of beef accounted for 22% of total global meat consumption in 2018, and it has increased by 1% per year in the past two decades. Australia accounted for around 3% of global beef production in 2017 and produced 2.3 million tons of beef and veal in 2018. The value of Australian beef exports in 2018 was $8.66 billion, up 16% from the previous year (Fig. 4.1). Australia exports about 70% of the beef it produces, which accounts for 15.37% of the world's beef exports, and it was the third largest beef exporter in 2018. In 2019, the value of Australian beef exports increased as well, which made Australia become the country with the highest dollar value worth of beef exports regardless of whether fresh, chilled, or frozen. These facts indicate the importance of the beef industry to Australia.

Over the last two decades, total global meat consumption has increased by 64% at an average rate of 1% for beef. The future demand for beef is likely to increase due to the growing global demand for protein. China's market has grown rapidly in the last couple of years due to rising incomes and the approval of more Australian processing facilities by the Chinese Government, from less than 8000 tons in 2009–2010 to 161,000 tons in 2013–14. In 2018, the world's largest importer of beef and veal (in volume terms) was China, overtaking the USA, with Japan as the third largest. Mainland China became Australia's fourth largest beef export market by volume (after Japan, the US, and South Korea), and accounted for 14.5% of Australia's total beef exports in 2018 (Fig. 4.2). The research by Cao et al. (2020)

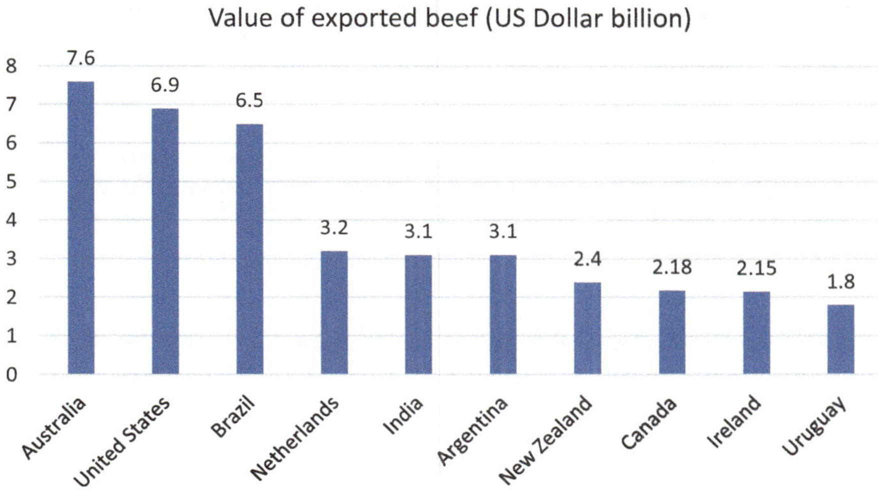

Fig. 4.1 Top ten countries that exported the highest dollar value worth of beef regardless of whether fresh, chilled, or frozen during 2019 (Source: Workman, 2020)

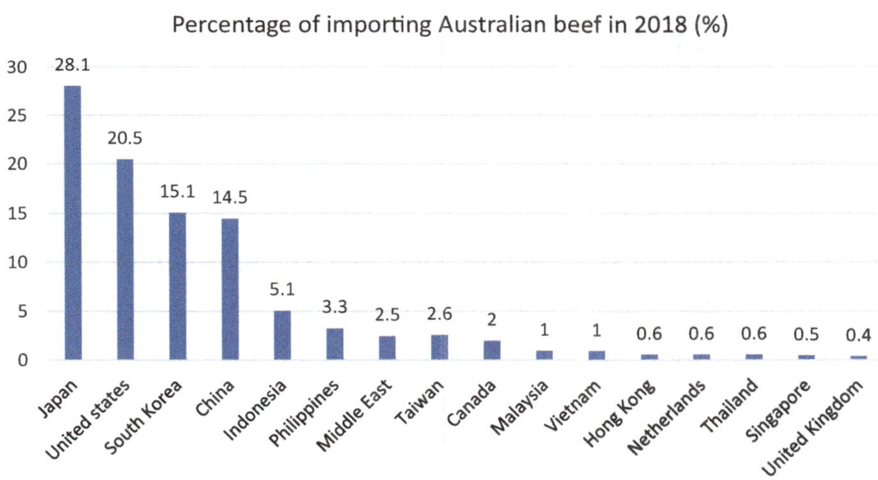

Fig. 4.2 Australia's beef export marekts' ranking by volume in 2018 (Source: MLA, 2019, p. 16)

also suggested that many Chinese consumers have the belief that Australian cut and packed beef products are better than Chinese-processed Australian beef products.

The Australian beef supply chain is one of the most complex; about 13 stakeholders are involved in one cross-border transaction (Fig. 4.3). It also has quality assurance standards and laws at every stage of the beef supply chain. From safety, labelling, and composition to food handling requirements, Australian food laws cover many possible food issues. There are three central elements of the red meat

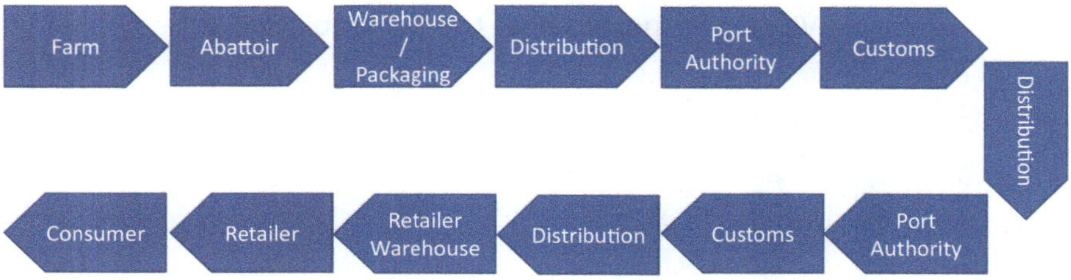

Fig. 4.3 Current model of cross-border supply chain (Source: Deloitte, 2020)

integrity systems: the National Livestock Identification System (NLIS); the Livestock Production Assurance Program (LPA); and the LPA National Vendor Declaration (LPA NVD). For quality performance, there is a Meat Standards Australia (MSA) for instance. The MSA index presents the eating quality potentials of each cut of the beef. The standards exist to improve meat quality consistency. Supply chain participants will be expected to meet all the standards. Beef products need to go through multiple processes and checks before going on the market.

4.2 Issues Within the Australian-China Beef Supply Chain

Through BeefLedger's internal research, the company identified the following issues in the cross-border beef supply chain.

4.2.1 Fragmented Information Flow and Misinformation Within the Cross-Border Supply Chain

The global supply chain is more complex than the domestic supply chain. The information flow is easily fragmented due to distance and the involvement of multiple actors. Information can be insufficient, incomplete, inaccurate, or inappropriate, which can lead to mistakes in product flow and finance flow. BeefLedger aims to address such misinformation, which includes hidden information and hidden action. For beef products, every second kilo of beef sold in China claiming to be Australian is not Australian. For Chinese consumers, the country of origin and temperature control during cold chain logistics are the most important pieces of information, as mentioned by Warwick. The country of origin can guarantee product authenticity and the temperature control can prove product quality. Such authentic information can enhance buyers' purchasing confidence. Warwick pointed out, *"the information about things is always part of what people buy. So, people never purchase things without at the same time also purchasing some information about them. It is not possible to deal with the world without some level of information. The existence of third parties such as government or NGOs can provide governance to the beef supply chain to some extent".*

Hidden actions can include replacing authentic products with fake products for huge profit. This can happen more often on the products with higher prices as there are higher motivations. For example, the retail price of BeefLedger's beef product is slightly higher than the base price due to the premium quality. To address the risks of hidden action and to prevent product fraud, strict regulations and laws are applied in the food industry, such as Australian meat and livestock industry regulations. According to Warwick, *"And both [hidden action and hidden information] of those are related to capricious information. So we have developed an infrastructure and a process that tackles these issues with information and we can do that whether it's one bit of information or whether it's a thousand bits of information, it doesn't matter".*

4.2.2 Inefficient Cross-Border Document and Payment Process

The exporting process is usually complicated with volumes of documents and signatures. Research conducted by IBM and Maersk suggested that it requires more than 200 interactions with 30 individuals and organisations, and a four-inch stack of paper records to complete a single shipment of goods from East Africa to Europe. Thus, IBM and Maersk worked together to digitalise the shipping process by blockchain technology. The long and complex exporting process is costly. BeefLedger researcher Dr. Cao pointed out, *"for exporting beef products, you need to provide all kinds of documents. There are many different checks are taken place in China border. These things are time consuming and human errors can easily happen".*

Moreover, cross-border payment is also an issue. The letter of credit is used in cross-border business to generate trust between business partners. However, the process of issuing a letter of credit is usually time-consuming. Meanwhile, the cross-border payment transaction usually takes a long time and is expensive.

4.2.3 Consumer Concerns

BeefLedger's original target customers are mainly Chinese customers, who care more about the country of origin and conditions of the products during shipping. The main reason to emphasise the importance of the origin is because food fraud is a massive issue in China, particularly in the case of imported expensive Australian beef steaks. Food fraud is an intentional deception to gain profit by using unethical behaviours such as mislabelling, substitution, and counterfeiting food products

(European Commission, 2020). The complex global food supply chain also makes it hard to detect food fraud. PwC (2020) proposed the following equation for food fraud vulnerability: opportunity + motivation—control measures. Food fraud usually happens in circumstances where the potential gain is high and the risk is low. For beef products, the large price difference between real and fake Australian beef products can be an incentive for fraudsters. The multiple tiers of the supply chain and the multi-mode of shipping between Australia and China also make food fraud detection more difficult. Thus, it would be difficult to track and trace back to where responsibility for the issue lies. Therefore, the authenticity of the beef products is one of the most important pieces of information for BeefLedger's customers.

The temperature during global shipping is another important piece of information for customers. Keeping track of the temperature of beef products during shipping is a way to ensure the safety of the product. For fresh products, the storage environment is very important, particularly during long-distance transportation. Temperature control is an important way to minimise the chances of harmful microbiological growth. For Chinese consumers, food safety is an important issue that is yet to be adequately addressed. Apart from the Sanlu toxic infant milk scandal in 2008, which caused the deaths of at least eight children, there are many other scandals that happen every year. All those food scandals cause Chinese customers to pay more attention to the food safety issue.

4.3 Blockchain Implementation

This section presents the process of blockchain implementation, including preparation, what has been done, and what has been left for the future (Table 4.1).

4.3.1 Preparation

Before launching its blockchain platform, BeefLedger carried out a year-long internal research and development initiative to learn the market better. The research and business mapping started in 2016, and the official project was launched in 2017. The internal research included three stages. The first stage was at the beginning of 2016; it mainly focused on collecting as much information as possible and identifying the potentials of blockchain for the beef supply chain. Warwick explained, *"The issue is whether or not there is business value for the supply chain and whether or not someone is willing to pay for which kinds of services".*

BeefLedger then moved to the second stage at the end of 2016 and formed BeefLedger Founders No. 1 Pty Ltd., an early-stage IP developer and curator company. In this stage, four to five colleagues were invited to co-invest in the project to undertake the next phase of the research and to map out the business scenario.

Table 4.1 An overview of different stages

Stage	Technology diffusion phases	Period	Stakeholders	Actions	Resources	Outcomes and expected outcomes
Preparation	Knowledge/ awareness Persuasion Decision	2016	BeefLedger Food Agility Queensland University of Technology (QUT)	1. Conducted a year-long internal research study 2. Business mapping 3. Formed a BeefLedger Founders No. 1 Pty Ltd	1. Information about blockchain 2. Investment from five colleagues.	2016 was a pre-feasibility period, BeefLedger carried out internal research to collect information about blockchain. BeefLedger had also considered business strategies at this stage.
Implementation	Implementation	2017 to present	BeefLedger IoT companies Food Agility Queensland University of Technology (QUT) LibertyPost Local agents	1. Blockchain project officially began (2017) 2. Food provenance pilot (2018) 3. Adopted RFID 4. Partnership with e-commerce LibertyPost and local distributors (2019) 5. Six city roadshow programs (2019) 6. Conducted research about consumers' value perceptions (2019) 7. POA blockchain platform ecosystem was completed (end 2019)	1. Various technology meetings. 2. Consumer review and feedback.	1. 2017 was a feasibility period, the platform was kicked off officially. 2. 2018 and 2019 were years of development. Teamed up with Food Agility, Queensland University of Technology (QUT), and truck insurer—NTI, BeefLedger had a pilot run in monitoring product provenance and supply chain process. 3. Moreover, some IoT devices were also adopted to increase data inputting accuracy and efficiency. 4. BeefLedger has also formed a partnership with e-commerce LibertyPost and local agents to sell the products. 5. Showcase and food tasting were helped to attract more consumers. 6. BeefLedger has also conducted new research about how consumers value a product.

(continued)

Table 4.1 (continued)

Stage	Technology diffusion phases	Period	Stakeholders	Actions	Resources	Outcomes and expected outcomes
						7. In 2019, the POA blockchain platform ecosystem was completed, BeefLedger has been moving to commercialise deployment since then.
Future	Confirmation/ continuation	In a near future	BeefLedger IoT companies Queensland University of Technology LibertyPost Local agents	1. Enhance consumer engagement 2. Provide a blockchain tool for other brands 3. Supply chain finance 4. Certification process	1. Experience and research reports from implementation stage.	1. BeefLedger is willing to engage more consumers and focus more on the consumer experience. 2. BeefLedger hopes to provide blockchain solutions to other companies, and to provide an industry standard. 3. Data on blockchain can also contribute to the supply chain finance such as insurance, financing, etc. 4. Blockchain can also streamline certification process, including paper reduction, time efficiency, security, etc.

4.3.2 Implementation

After the study, BeefLedger decided to kick off the implementation on beef traceability. The last stage was to officially set up BeefLedger as a blockchain company in late 2017. *"We do internal pre-feasibility, then we begin. So, 2016 was pre-feasibility, 2017 was feasibility and then 2018 and 2019 has been development"*, said Warwick. In December 2018, the Food Agility and Queensland University of Technology (QUT) took part in the BeefLedger food provenance pilot to track the Australia-China cross-border beef supply chain. The pilot was conducted to improve food safety, animal welfare, and export security. At this time, the platform allowed supply chain participants to create their own unique identity and to start uploading information on the chain. Supply chain partners followed seven steps to take part in this initiative: (1) Create company account; (2) Account confirmation; (3) Log-in account; (4) Set up restrictions; (5) Verify data through multi-signature process; (6) Add company members and grant permission; and (7) Eligible members can update real-time information (produce information, pick-up information, drop-off information, measure information).

Warwick said, *"When you are using blockchain technologies as a ledger structure to interface with the real world, you need to have ways in which the blockchain architecture relates to the external environment. And so you need to have different kinds of informational interfaces at the production or at the collection"*. To best engage data collection, BeefLedger works with IoT device manufacturers who specialise in asset identification. A few IoT devices were tested to also find the best data collection solution. BeefLedger first tested the near-field communication technology chip, which works effectively for end consumers. However, it is not practical as it did not support cost-effective bulk data uploading, which is necessary in industrial production settings. In this case, BeefLedger adopted Radio Frequency Identity (RFID) to receive data. For livestock, an industry-standard RFID tag is attached to an animal's ear. Newer so-called smart tags are also in development from many innovators, which enable the ability to monitor animals' health such as temperature, heart rates, and blood oxygen levels. Animal identification via standard RFID ear tags can be scanned with either a handheld scanning wand or by a fixed RFID reader, with data sent to the local server. It can greatly enhance data accuracy and data collection speed while minimising human error. RFID labels are also used on packaging of the finished product to enable effective identification. However, RFID also has a drawback relating to its high cost. Therefore, BeefLedger is also looking for a more cost-efficient data collection device to replace RFID.

After interface testing, BeefLedger provided a bridge to link the blockchain database and consumers with a QR code. It is an easy way for consumers to reach traceability information since consumer cannot access RFID scanning. Once the information is uploaded and verified by multiple stakeholders, consumers can obtain traceability information by simply scanning the QR code. The information for consumers usually includes text, pictures of export and import documents, and a one-minute video. The short video introduces the living environment of the animals, producers, the people in the community, and other relevant details.

BeefLedger also actively participated in many meetings to connect with experts in the industry and held a few food-tasting and showcase events to learn more about consumers' preferences. In 2019, six city roadshow programs were held to present consumers with a range of premium Australian beef products. Working with e-commerce LibertyPost and local distributors also enabled three different operation models to be tested: (i) direct business-to-consumer (B2C) based on individual purchases, (ii) direct B2C based on group buying, and (iii) downstream franchise support by activating social networks (economy of acquaintances). This can improve real-time information flow within the supply chain and creates a feedback-loop between consumers and producers. Dr. Cao suggested, *"It can form interaction. For example, consumers can know the production process, and the producer can receive consumer feedback"*. Warwick further said: *"Supply chains are actually changed from chains to cycles; they are circuits, the economic circuits. So, every stage in the circuit has its own need for information and players who are engaged in each of those stages will perform different roles. The farmer can be validated in the network for information that is collected at a different point in the process. Nothing stopped somebody participating as a validator in the network. Nothing stops somebody running a storage node to store a copy of the information for the whole network"*. Moreover, in late 2019, Dr. Cao led research about consumers' value perceptions on Australian beef products. The research suggested that consumers are willing to pay a premium price for products with blockchain-enabled traceability. However, *"willingness to pay is very different from actual behaviour"*, said Warwick.

2020 brought unprecedented challenges and opportunities to BeefLedger, due to the effects of COVID-19 and the changing geopolitical landscape. The COVID-19 pandemic impacted consumer behaviours and distribution channels. Prior to COVID-19 the strongest channels for imported meat were supermarkets and food services. The lockdown significantly impaired the food services sector and saw a boost to the e-commerce sector. Consumers shifted to home delivery models to access their life necessities. Consumers have also become somewhat less "extravagant" in their daily spending habits.

In response, BeefLedger accelerated plans to shift to a more intensive B2C footing, with direct delivery to household consumers or stronger efforts to support localised distribution partnerships. New business models were required, which involved streamlining the logistics dimensions of supply chains and activating "group buying" models for more direct at-scale engagement with consumer communities. This adaptive process in effect focuses on shortening supply chains, removing rent-seeking layers, and delivering cost-competitive products to consumers. COVID-19 has also increased consumer (and regulator) concerns about supply chain security and safety. There is now more interest in the role of digital technologies in enabling more effective supply chain supervision and management.

In terms of the impact of geopolitical dynamics, the direct implications are difficult to determine as consumer interest in Australian beef remains high.

However, from BeefLedger's perspective, opportunities to support the delivery of blockchain-credentialed product of excellence into the China market from other beef-producing regions of the world have beckoned. During 2020, BeefLedger was contacted by teams in South Africa and South America, to explore cooperation on technology deployment and market access opportunities. A joint venture has now been established in South Africa (with mainly a domestic focus at present), and work continues with colleagues in South America to fine-tune an approach to export channels. In light of changing consumer sensitivities, opportunities to support the supply of blockchain-credentialled beef at good prices is being actively explored.

4.3.3 Coordinating with Stakeholders

For the stakeholders who participate in the blockchain system, the information uploaded to blockchain will need to go through a multi-signatory process. The information needs to be validated by multiple actors to ensure its validity. This not only saves stakeholders from relying on third partiers—NGOs and governments—but also encourages stakeholders to actively take responsibilities for checking and uploading authentic data. Once the data are uploaded and validated successfully, the record is permanently on blockchain. The record can be traced at any time by anyone on the blockchain. It is also a reminder for stakeholders to pay attention while uploading data.

Tokenisation can encourage legitimate behaviours of stakeholders within the supply chain. After scanning the animal ID tag, an identity token is created. The token is also an asset ownership certificate, which is uniquely connected to the animal's identity. In the retailing stage, the product should be sold with the specific token in order to prove the authenticity of the meat, because the unique identity token can show the traceability of the item's custodian/owner. The tokenisation makes tracing and tracking of responsible parties much easier. This not only brings transparency and convenience to stakeholders for traceability purposes but also drives the behaviours of the stakeholders towards more legitimate and ethical trade.

In some circumstances, when the final products are associated with a unique token, even though the real product can still be replaced by a fake or cheap product, the value of the real product will necessarily drop due to the lack of an identity token. There are four scenarios (Table 4.2) that can show the equation of the risks and rewards, and how the economics can drive people's behaviours. Scenario 1 is when the real product with a token and the cheap product can both realise £50 in profit. Scenario 2 is when the products are swapped. In this case, the real product makes zero profit, while the cheap product makes £100 profit. There is no change to the total profits of the two scenarios. Scenarios 3 and 4 show the same principle when beef recorded on blockchain can sell at a premium. The four scenarios suggest that the fake beef will not be able to make extra profit but does take higher risks. In this case, there is no reward for players taking risks in swapping products. As mentioned above, the equation of food fraud vulnerability is opportunity + motivation—control

Table 4.2 Four scenarios

	Scenario 1		Scenario 2	
	Real (with token)	Cheap/fake	Real	Cheap/fake (with token)
Cost (£)	100	50	100	50
Market price (£)	150	100	100	150
Profit (£)	50	50	0	100
Overall profit (£)	100		100	
	Scenario 3		Scenario 4	
	Real (with token)	Cheap/fake	Real	Cheap/fake (with token)
Cost (£)	100	50	100	50
Market price (£)	250	100	100	250
Profit (£)	150	50	0	200
Overall profit (£)	200		200	

measures (PwC, 2020). There is no motivation to swap products for zero profit. Therefore, blockchain is not a tool that prevents people from swapping products; instead, it creates economic incentives to align behaviours. Warwick explained, *"This is what I always say that what blockchain can do is bring two different worlds together. Traditionally, we have the world of money and we have the world of anti-counterfeiting. The money world, just money. The anti-counterfeiting world is all about regulations and policing. By using economics to change the incentive mechanisms and the transparency of information, we can actually drive behaviour more and more towards the legitimate trade and weed out the illegitimate trade"*.

Apart from the stakeholders within the supply chain, blockchain can also encourage other practitioners in the market to use blockchain to stay competitive. The economics behind this is that between similar products, the value of the one that does not provide additional features will drop. More specifically, the prices of two similar pieces of meat are the same before using blockchain. When BeefLedger is using blockchain to provide information transparency, the value of similar but not blockchain-tracked meat will drop. This is due to the fact that information is always a part of what people buy. Research by Deloitte (2020) shows that 37% of consumers would be willing to change to brands that share more product information. Rather than buying meat itself, customers also pay for the information on the meat such as brand and originality. As Warwick said, *"The information about things is always part of what people buy. People never purchase things without at the same time also purchasing some information about things"*. In this case, the blockchain can provide more information to assist customers' purchasing decisions. For two similar products with the same price, customers tend to choose the one with extra benefits or additional features, including additional information. It is also an encouragement for all practitioners in the market to improve their service and provide customers with better information (Pazaitis et al., 2017).

For customers, blockchain applications can address misinformation and enhance their purchasing confidence. In many cases, customers can be misled by information or labels particularly for imported products that are hard to track. From the consumer perception study by Grunert (2005), quality labels are not powerful incentives for customers. This means that customers do not feel that the quality labels can fully present the products' quality and have little confidence in using them as buying cues. As mentioned above, the country of origin and the temperature of beef steaks are the main pieces of information for Chinese customers in the case of BeefLedger. Food fraud and food safety are huge issues in China. By using blockchain, customers can acquire quality and authentic information to ensure the quality of the products. The transparency of the information can largely increase purchasing confidence.

4.3.4 Future Plan

The POA blockchain platform ecosystem was completed by the end of 2019, and BeefLedger has been moving to commercialise deployment since then. In the future, there will be some improvements to both the platform and the business strategy. On the platform perspective, BeefLedger hopes to focus more on the consumer side. *"Most of our work, we don't necessarily go all the way to the end consumer yet. So, our focus has been mainly on the wholesale buyers at this stage, and the next version or the version after. We will look at the end consumer in a lot more detail"*, said Warwick. The report from Deloitte also suggests that consumer engagement and the consumer feedback loop are the future direction. Consumers will have more confidence in their purchase decisions. The more accurate consumer feedback will also encourage stakeholders' engagement.

From the business strategy perspective, the blockchain-enabled traceability platform for the beef supply chain is only a first step, Warwick claimed: *"Our business strategy is to support other brands as a blockchain solution provider rather than build just the BeefLedger beef brand. So if you think of a computer that has many different brands, but remember on the computer there's often the little sticker that says Intel inside, we would like to be Intel inside"*. BeefLedger hopes to attract other brands to use its platform to track products, and gradually help the beef industry—and even the wider food industry to build tracking standards. Dr. Cao claimed, *"Blockchain-enabled traceability can become an industry standard, and all food products will be traceable"*. Warwick added, *"The market will move quicker than that. Social networks and social e-Commerce and social media make the markets work a lot quicker than they ever did. So, consumers will move incredibly fast, in the face of getting something for nothing. So, and you know, more and more things become default. So, over time, more information layers will become default layers based on a relationship of value"*.

Apart from bringing transparency and security to the food supply chain, the data collected on blockchain can generate more value, such as precision marketing and supply chain financing. Just like Dr. Cao said, *"from an industrial application point of view, it is equivalent to having industry data, and then you can do a lot of things, such as financing, insurance and others"*. The real-information flow can also assist companies in making evidence-based decisions. Blockchain can not only be used as a tool for traceability; there is a bigger value behind it.

Moreover, certification on blockchain is another future direction. Certifications can be stored, managed, and shared on blockchain. This can significantly improve the efficiency of the quality accreditation process. Paper and time can be reduced as data are digitalised and shared on blockchain. The process of certification issuance is also secured.

4.4 Blockchain Benefits

This section presents the benefits that blockchain can bring to BeefLedger and how can it solve the issues mentioned above. The benefits include supply chain information flow optimisation, streamlining payment and documentation, and enhancing consumer buying confidence.

4.4.1 Optimising the Information Flow for BeefLedger

By using blockchain, a more efficient information flow can be achieved. More specifically, traditionally, the information passes from actor to actor within the supply chain from the beginning of a supply chain to the end consumers. The information system is an isolated communication system. In this case, raw material suppliers have no idea about the information about the final sales or customer reviews, as there is no direct contact between them. This is because the actors in a supply chain are usually only in contact with their direct business partners. This can lead to poor communication within a supply chain, particularly when it comes to the global food supply chain, which is usually more complex. The isolated information also leads to the bullwhip effect, which describes a phenomenon of a growing variation upstream within a supply chain. It is normally due to vague information which causes order mistakes between final sales and the original producer.

In the context of BeefLedger, due to the decentralisation feature of the blockchain, the supply chain changes from a linear line to a circuit (Figs. 4.4 and 4.5). There are more than 15 stakeholders and eight key stakeholders in the supply

Fig. 4.4 Linear supply chain (Source: the author)

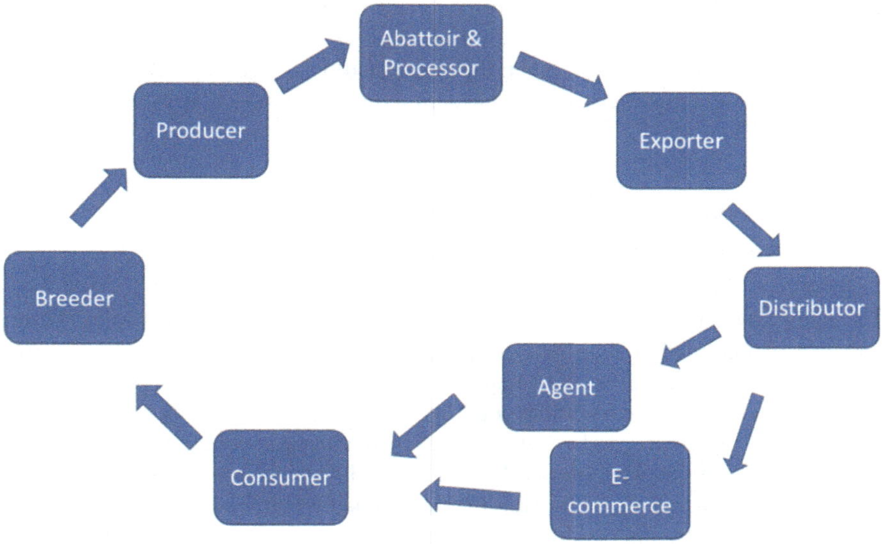

Fig. 4.5 Circulate supply chain (Source: the author)

chain of BeefLedger; these include breeders, producers, transport companies, packaging companies and exporters, among others. Every actor can engage with the blockchain system and share the same information at the same time. Not only can the speed of information sharing be greatly improved, but the information asymmetry among stakeholders can also be avoided. In the case of BeefLedger, for example, producers can upload information about the animals and help to verify data that are collected from breeders. Meanwhile, producers can also go to the end of the supply chain to check the sales data and receive feedback of the products from consumers in order to facilitate operational decision-making (e.g., which type of beef meat is more popular). As Dr. Cao suggested, *"from the process perspective, it can form a kind of interaction. For example, as a producer, he can understand the entire process of his product. In addition, he can understand the feedback from consumers, and to form an interactive mode. In many cases, Australia's traditional supply chain mode is fragmented in this way, and farmers receive insufficient and processed information. The information inconsistent and lagging with the market trend. Therefore, it is important to achieve real time Information flow, it will also help evidence-based decision making"*. Therefore, not only is a more connected network with better peer involvement built through blockchain but a better coordination between stakeholders is also developed.

Besides, digitalised information on blockchain is much easier to access, check, store, and trace than paper documents, which will largely improve information sharing and reduce information asymmetry. Moreover, blockchain can create data dependability and reliability for BeefLedger. Blockchain is more than just uploading data, it is also important to collectively verify these data. The multi-signatory process can make information processing more efficient, as there is no need for third parties or siloed repetition/duplication. The information can only be allowed to upload to the blockchain once multiple others have finished data-verifying. This means that data quality and authenticity can be ensured at the same time as data inputting.

4.4.2 Streamlining Cross-Border Documentation and Payment

By using smart contracts, some processes can be executed automatically, so the strength of automation is emphasised (Babich & Hilary, 2018). This will bring significant convenience to the supply chain. Rather than paper-based, a smart contract is a set of self-executable code, which will speed up the processing speed. For instance, payments can be executed automatically once certain data state conditions are met. In the case of BeefLedger, cross-border payments are much more secure and easier to process by smart contract-enabled escrow and spot payments. Meanwhile, tokenisation combined with the smart contract can achieve better supply chain efficiency and data security. In the BeefLedger context, a unique token is an ownership of the beef product. It is created after scanning an animal ID tag. The token will add information by smart contract and become a new token following the events (e.g. processing, exporting) that typically take place in a supply chain. Therefore, a beef product always has its identity token with the information to prove its authenticity.

4.4.3 Enhance Consumer Buying Confidence

The feature of decentralisation can provide stakeholders a trust-free environment within a supply chain and save them from having to rely on third parties. Stakeholders can make deals without trusting each other. Although the traditional centralised supply chain has many third parties such as governments and international groups that provide regular checks and feedback on product quality, this does not exclude the possibility of undetected dishonest behaviours and bribery. Ironically, the Chinese local authorities were under investigation due to bribery in the toxic milk scandal in 2008. However, by using blockchain, the information needs to through a multi-signatory process to be verified to make sure it is authentic. The multi-signatory process can bring the strength of data validation, aggregation, and visibility (Babich & Hilary, 2018). The data will come from multiple sources, but also become validated while they are uploaded by the multiple stakeholders along the supply chain. As data from every involved stakeholder are shared along the supply chain due to decentralisation, data visibility can also be ensured. In this case, there is no need for either a central power or third parties.

Blockchain's feature of immutability can ensure historical data integrity and authenticity. Any data changes will be recorded, which will prevent dishonest actions such as data tampering. Traditionally, information is usually controlled by some large central companies that can escape their responsibilities by altering or erasing records. For example, Volkswagen admitted cheating emission tests by changing the CO_2 emissions level to cope with the relevant tests (Hotten, 2015). Once data are verified and uploaded to the blockchain, the records become permanent, which also provides a more efficient traceability of the products. During food recall, efficient traceability can be a valuable resource to trace and find the origins of the contaminated food. Immutability is a key enabler for information fraud resistance. Meanwhile, data are replicated on every node, which can ensure data resiliency. In turn, this will prevent a single failure from causing the failure of the whole database.

4.5 Implementation Barriers

There are four perspectives of barriers according to Saberi et al. (2018): intra-organisational, inter-organisational, systems-related, and external. This section presents how BeefLedger overcame the barriers it faced (Table 4.3).

With the rich experience and deep insights into the beef industry, Warwick and his business partners founded BeefLedger with goals to provide authentic products to consumers. The aims to address issues in global beef supply chain and to provide credibility to consumers are strong driving forces to BeefLedger. After a few research and test runs, the BeefLedger project was officially launched in 2017. The beef products that are blockchain-traceable are in the middle- to high-end price range. At this price range, consumers tend to care more about food quality and have high requirements on food traceability.

The blockchain-enabled platform will require multiple stakeholders to participate. BeefLedger's multi-tier suppliers have little knowledge and understanding of blockchain. Meanwhile, their confidence in and concerns about using blockchain are also different, including data privacy, implementing costs, and possible changes to the company operations. BeefLedger employs the community value department and economic department to provide education and economic incentives to its stakeholders. To make sure all stakeholders are on board, it is important to show that blockchain implementation is not only for customers, but also benefits all the stakeholders within the supply chain. For retailers, tokenisation and data immutability can ensure the authentic value of the products. For producers, the combination of smart contract and IOT can bring much more convenience and accuracy in data inputting.

Moreover, adopting blockchain is easy and cost-efficient. It will not increase the workload and costs, nor is there a need to change the existing system. Data privacy can be guaranteed by the blockchain governance structure, and companies get to choose the information and the stakeholders they are willing to share with. Apart

Table 4.3 Challenges and overcome methods

Four perspectives	Barriers' categories	Barriers in practice	Approaches and solutions	Supporting Quotations
Intra-organisational barriers: the barriers in this category is from the internal perspective of the organisation	Company demand and determination	Challenge in finding company demand before adopting technology and keeping strong determination during blockchain implementation.	BeefLedger started the project with a product in the middle to high-end price range and faces a high possibility of counterfeiting, which means that consumers care a great deal about the product quality and authenticity.	"There is a of quality. There is a range. And I think you would say that the starting position for Australian beef is in the middle market and then it goes all the way to the luxury market. Okay. So, from a pricing perspective, for example the beef could be from 25 or 28 USD, per kilo all the way to hundreds of USD per kilo. The issue is whether or not there is value for the supply chain and whether or not someone is willing to pay for different kinds of services. So, of course the more expensive something is and the greater the risk that that product could be counterfeited the more people will pay for improved validity and certainty of authenticity, but you can also implement very low-cost solutions or individual fruit items. What people need to think about is the blockchain as

(continued)

Table 4.3 (continued)

Four perspectives	Barriers' categories	Barriers in practice	Approaches and solutions	Supporting Quotations
				an infrastructure is very flexible. How much information is stored through a blockchain process really depends on what the needs of the industry are. So, for a very expensive product, it is likely that consumers will demand a lot more information and therefore the cost of collecting or making that information is a lot higher than otherwise".—Warwick Powell, chairman of BeefLedger "Like Warwick and Charles, they are familiar with the beef industry, and they have resources in it. They discovered these problems in the industry through this way, and that there was just such an opportunity. The demand for Australian beef is particularly large. In fact, another reason, from the perspective of the Australian industry, because Chinese Australian beef is equivalent

(continued)

Table 4.3 (continued)

Four perspectives	Barriers' categories	Barriers in practice	Approaches and solutions	Supporting Quotations
				to the highest quality in China, it is also because of high possibility of fraud product, blockchain can be used strengthen a brand protection." —Dr. Cao, BeefLedger researcher
			BeefLedger had a long-term preparation before kicking off the project, including a year of internet research and business mapping.	"We did about a year of internal research and development before launching the project. So we officially launched the project around October, November, 2017. So we began our internal research and development and business case mapping in 2016". —Warwick Powell, Chairman of BeefLedger
	Financial constraints	Massive initial investment	To conduct the blockchain project, BeefLedger not only invited private investors, but also received financial support from the government.	"In fact, our project is a business plus university plus industry, and then project funds come in. For example, if we do this project, because this project is an Australian funding organization, they also donate part of the funds, and then BeefLedger also provides part of

(continued)

Table 4.3 (continued)

Four perspectives	Barriers' categories	Barriers in practice	Approaches and solutions	Supporting Quotations
				the funds, and then it is equivalent to a project with a total value. It is not an operation of a pure enterprise, it is a kind of cooperation that is equivalent to enterprise + industry + funding organisation, which means co-promotion".—Dr. Cao, BeefLedger researcher
Inter Inter-organisational barriers: the barriers are between the supply chain stakeholders	Stakeholder engagement	Challenge in motivating stakeholders to join	BeefLedger has invited a professional to educate stakeholders and presented the values and the benefits to different stakeholders.	"We are now equivalent to gradual advancement, and first include qualified upstream producers to come in. And through the infrastructure investment of the Australian government, it is equivalent to a step-by-step process to start to include those producers in other remote areas. Moreover, we have a professional who is checking producers' acceptance from a cultural perspective, and then educating the producer through the market perspective. For example, by

(continued)

Table 4.3 (continued)

Four perspectives	Barriers' categories	Barriers in practice	Approaches and solutions	Supporting Quotations
				discovering consumer data and feedback, that means that after the producer enters our ecosystem, they can view consumer feedback and understand dynamics of your product. In addition, if some companies think this thing is great, they may recommend some other people to use it through the recommendation system".—Dr. Cao, BeefLedger researcher "It's about sustaining a competitive position because in the end, if they don't do it, they will be the ones that get discounted. The market will move quicker than that. Social networks and social eCommerce and social media make the markets work a lot quicker than they ever did. So consumers will move incredibly fast. More and more things become default. So over time, more information layers will become default

(continued)

Table 4.3 (continued)

Four perspectives	Barriers' categories	Barriers in practice	Approaches and solutions	Supporting Quotations
				layers based on a relationship of value, what people want to know and costs, how much they are willing to pay for it."—Warwick Powell, Chairman of BeefLedger
			Lead by example is also what BeefLedger is trying to do. So far, a few stakeholders participated in the beef blockchain, including producer, abattoir (exporter), and transport company.	"Let's say we use our beef supply chain as an example to let others find blockchain is indeed useful. They can get more consumer information and understand the consumer's behaviour to promote their products. We hope they can see the value and will join us gradually".—Dr. Cao, BeefLedger researcher
			Perceived ease of use is also a key factor to impact stakeholder decisions. Therefore, BeefLedger has to make the system easy to use and can fit in different systems.	"Two considerations before companies decide to adopt blockchain. First, after I use your traceability system, how much does it cost to my business? The second question is if they have to change the operation of the existing system. From the perspective of the

(continued)

Table 4.3 (continued)

Four perspectives	Barriers' categories	Barriers in practice	Approaches and solutions	Supporting Quotations
				business process, it hasn't changed much. It's just equivalent to scanning the code after products arrive, or manually checking it, uploading the data, and storing it in the warehouse. It is equivalent to adding an information confirmation in each step. In fact, companies have to confirm products and data anyway. It is usually a manual confirmation, they may just record the information in a notebook, or have an excel sheet after the manual confirmation."—Dr. Cao, BeefLedger researcher
		Financial pressure for stakeholders to adopt a new technology	BeefLedger calculated the cost and suggested that there will not be many costs apart from those for IoT devices. The system itself will be affordable.	"We are developing this system to obtain data. Basically, now we have adopted the RFID. The fact is that this thing costs a small fixed asset after you buy it. The starting price of that item is quite high, but this item is not too cheap no matter where you buy it (if the Company has IoT

(continued)

Table 4.3 (continued)

Four perspectives	Barriers' categories	Barriers in practice	Approaches and solutions	Supporting Quotations
				devices already). If that is the case, it is equivalent to very low cost. We have calculated internally. It is basically equivalent to adding a package and a scan code for each product, and uploading the information to the chain. In fact, one package for a product may cost one or two cents".—Dr. Cao, BeefLedger researcher
		Concern in data privacy	Stakeholders may have concerns on their data privacy. To solve this issue, BeefLedger conducted a governance structure both on-chain and off-chain. Companies can decide what kind of information they are willing to share, and who gets to see what information.	"If you want everything on the chain, you have to have an on-chain and off-chain governance structure. This involves a problem of governance structure on the chain. For example, some information may be that companies are willing to publish, and some information is that companies are unwilling to publish, that is, they can only say what they just like on our platform, we have also designed a group's permissions. As a participant, you can set which

(continued)

Table 4.3 (continued)

Four perspectives	Barriers' categories	Barriers in practice	Approaches and solutions	Supporting Quotations
				information you want publish to the entire chain for others to see, and which may even have many protections. You can say which information you can let upstream people see, which information can be seen by people in the next step, and which information can only be seen by some executives of the Company".—Dr. Cao, BeefLedger researcher "You don't need to tell people the price of something to give them information that provides people confidence in the authenticity of the product. No one's demanding that LV reveals its costs other than through the annual reports that it does as a public company".—Warwick Powell, Chairman of BeefLedger
		Consumer engagement	From the perspective of consumer awareness, customers need to trust the process, the concept, and the specific	"How much information is stored through a blockchain process really depends on what the needs of the

(continued)

Table 4.3 (continued)

Four perspectives	Barriers' categories	Barriers in practice	Approaches and solutions	Supporting Quotations
			implementation, even they have little understanding of blockchain. Most possibly, customers will get used to blockchain, and trust on the products with labels written blockchain traceable in the end.	industry is. So for a very expensive product, it is likely that consumers will demand a lot more information and therefore the cost of collecting or making that information is a lot higher than otherwise. Consumers usually have what I would call background knowledge, which is just a sense of the value of things because other people talk about that, they hear it on the news, they see companies talk about things, and over time certain things become natural and that somebody doesn't know a lot about it at all."—Warwick Powell, Chairman of BeefLedger "We also are now developing now mechanisms by which the wider community of people also become an active participant in the network as consumers. So, typically, supply chain blockchain design is always

(continued)

Table 4.3 (continued)

Four perspectives	Barriers' categories	Barriers in practice	Approaches and solutions	Supporting Quotations
				focused on the actors who produce and transport products. We are now moving to a version that will allow end consumers to become participants in this blockchain network".—Warwick Powell, Chairman of BeefLedger
			Cultural difference between BeefLedger and Chinese consumers. To solve this issue, BeefLedger had conducted consumers' perceptions of value in 2019 to understand consumer behaviours.	"In fact, the thinking modes of Chinese consumers and the Western consumers are completely different. For example, if you design a kind of social software or something similar based on the consumption mode and consumption behaviour of the Westerns, the Chinese consumers will feel like it is impossible to use it. But if you take it to the Western world, it will be accepted. If you get something that is very good in Australia and bring it to China, Chinese consumers may not want to use it."—Dr. Cao,

(continued)

Table 4.3 (continued)

Four perspectives	Barriers' categories	Barriers in practice	Approaches and solutions	Supporting Quotations
				BeefLedger researcher
		Challenge in providing a good user experience	BeefLedger has tested multiple interface designs to find a more customer-friendly way for consumers.	"One of the challenges is being able to make this technology as easy for people to use as possible. And the technology, the blockchain industry itself I think understands that that's probably one of its weakest parts is that it is still not as easy to use and as easy to interface with as the more conventional information system is improving. So, it's . . . we've done some research in November as well on some consumers so. . . but we're testing a range of interfaces. Obviously the simplest one is through the QR code. We've been improving the app interfaces for users, whereas before our focus on development work was on the database backend systems. So, I would think that in the next two years easily that they will eat many APPs that will be very easy

(continued)

Table 4.3 (continued)

Four perspectives	Barriers' categories	Barriers in practice	Approaches and solutions	Supporting Quotations
				to use that actually be blockchain-capable or integrated to blockchain. But people won't know the difference".—Warwick Powell, Chairman of BeefLedger "Blockchain authentication means that there is a lot of information to upload after authentication. We are building a kind of knowledge management, which classifies different knowledge. After we provide several types of stratification, consumers may just look at this thing and know that it is true, without any further understanding. But some people may just want to know more, so they look for the second level of knowledge."—Dr. Cao, BeefLedger researcher
Systems-related barriers: the barriers relate to the technology	Technology immaturity	Scalability issues	Product information from breeder to the end consumers can be a great amount, including text,	"In fact, the challenge we are facing now is the scale of the Ethereum platform we use now,

(continued)

Table 4.3 (continued)

Four perspectives	Barriers' categories	Barriers in practice	Approaches and solutions	Supporting Quotations
			pictures and videos which require a large storage space and high transaction speed. Data storage and scalability need to be solved along with BeefLedger's growth. Currently, moving data to Holo chain or BSN are all under consideration.	how can we extend its scalability to carry large transaction volume in the future? Its infrastructure is in the process of slow development and has not yet achieved such a strong transaction speed or carrying capacity. Some people are not talking about using cloud storage. You saw it—consumers can view not just text but also some short videos. For things like this, if you produce every transaction, in the end, it means that the data storage is huge at that time. One of our staff has mentioned Holo chain. Then it means that some of these problems may be solved by Holo chain. At that time, he said that it is equivalent to establishing a regional database, we also want to use the BSN in China, to link or upload a part of data on there".—Dr. Cao, BeefLedger researcher

(continued)

Table 4.3 (continued)

Four perspectives	Barriers' categories	Barriers in practice	Approaches and solutions	Supporting Quotations
		Limited functions	Blockchain itself cannot do everything. A combination of blockchain and other devices may achieve a better efficiency.	"But when you are using blockchain technologies as a ledger structure to interface with the real world, you need to have ways in which the blockchain architecture relates to the external environment. And so you need to have different kinds of informational interfaces at the production in or at the collection. And internet of things, data is one, but basically you just need computers".—Warwick Powell, Chairman of BeefLedger
		The "garbage in, garbage out" problem	Deploying IoT devices and working with third parties can potentially solve the "garbage in, garbage out" problem.	"And I scanned through either a handheld scanning wand or they scan when the cattle come into what they call a crash, the catalogue held still. And there's a scanner, on the side of the crush."—Warwick Powell, Chairman of BeefLedger "We work with device manufacturers who specialise in temperature gauges, who

(continued)

Table 4.3 (continued)

Four perspectives	Barriers' categories	Barriers in practice	Approaches and solutions	Supporting Quotations
				specialise in geospatial tracking. They specialise in asset identification such as RFID asset identities who provide technology that enable us to collect information from vehicles. So that we have IBM. It's, so that's in vehicle monitoring systems. We work with people who have developed devices that are attached to the animals to measure our animal temperatures and animal heart rates and blood oxygen levels so that we could monitor animal health and conditions. So we do need to, and we do work with people who are involved in many specialised areas".—Warwick Powell, Chairman of BeefLedger
			BeefLedger has designed a multi-signature system to verify data on blockchain. To upload data, multiple verifications from different users are required; this can	"We have also implemented a multi signatory process by which data are committed to the blockchain itself, which means that we ensure that participants have

(continued)

Table 4.3 (continued)

Four perspectives	Barriers' categories	Barriers in practice	Approaches and solutions	Supporting Quotations
			significantly eliminate data mistakes.	to take active responsibility for checking the validity of information that they submit. And that you need to have more than one person agree that the information is valid before it can go onto the blockchain".—Warwick Powell, Chairman of BeefLedger "You use sensors to obtain data, after you obtain them, you need someone to verify them. Retailers want to trace the source of responsibility. For example, to trace who is responsible for the product in each link in the process, if something goes wrong, it can be directly isolated".—Dr. Cao, BeefLedger researcher
	Environmental restrictions	Some rural areas suffer limited digitalisation such as unstable network connectivity.	Develop a data interface that works well for basic CSV and PDF files and improve the local digitalisation where necessary.	"Instead of trying to change all of that in one go, we designed data interfaces that would actually just use CSV files. We use PDF files as event evidence. We don't change them, we submit them and they get stored

(continued)

Table 4.3 (continued)

Four perspectives	Barriers' categories	Barriers in practice	Approaches and solutions	Supporting Quotations
				and registered to the blockchain at PDF box. That's just reality. So we just took a very practical view, which is that what people most want to know, and what is the quickest way that we could give them the assurance that the information was more likely to be accurate than without trying to create a totally new digital infrastructure. Now we are progressively working with people in industry to improve digitalisation only where it is cost-effective to do so".—Warwick Powell, Chairman of BeefLedger
	Lack of skilled talent	Most software workers have limited industrial experience.	BeefLedger had to provide a certain level of industrial training to the software workers for a better cooperation.	"It is just a general challenge with a lot of software, you've got to educate them about what it is that you need to do and what the operating situation. It is just about building a bridge between software developers and the needs of the real world".—

(continued)

Table 4.3 (continued)

Four perspectives	Barriers' categories	Barriers in practice	Approaches and solutions	Supporting Quotations
				Warwick Powell, Chairman of BeefLedger
External barriers: the barriers from external environment.	Issues relate to policies and regulations	Regulation issues	To overcome the China firewall, it is necessary to transfer information and store it on a server in China. A legal department is also built for regulation issues.	"The barrier is actually coming in from outside China into China. So, it is really a one-way road with a firewall, so that has created the need to have what we call cross-chain solution. So, you've got to be able to cross the firewall. We then have to take the information and then transfer it into China and store it on a China site server. So actually, at this stage, we lose an automatic link to the blockchain."—Warwick Powell, Chairman of BeefLedger "Because there is a new legislation in China that requires that China's data must be stored in China, so now we are also considering establishing one server in China, one in Australia."—Dr. Cao, BeefLedger researcher
			Data protection laws can restrict BeefLedger from uploading and	"We have a team that specialises in this area, which involves

(continued)

Table 4.3 (continued)

Four perspectives	Barriers' categories	Barriers in practice	Approaches and solutions	Supporting Quotations
			collecting certain information. Therefore, BeefLedger has a dedicated team to target the issues in legal area.	legislation, legal regulation, and how to operate legally and compliantly. There must be a lot of considerations for this blockchain. We have dedicated human resources in this area to solve problems of how to put contracts on the chain, and legal compliance for data supervision and monitoring. Because there is a lot of legislation and cyber security, and it involves cross-border data exchange, cryptography and other things".—Dr. Cao, BeefLedger researcher
		Intense international relationships	In recent years, the intense international relationship between China and Australia has caused unstable policies. BeefLedger believes that such intensity is only temporary, and holds a positive attitude to it.	"In fact, from the overall perspective, I feel that this volatile relationship between China and Australia has a certain impact on the overall economic situation. But we should not stop the cooperation between China and Australia because of the current issue. In fact, there are many cooperative

(continued)

Table 4.3 (continued)

Four perspectives	Barriers' categories	Barriers in practice	Approaches and solutions	Supporting Quotations
				projects between China and Australia, and even some things like free trade. In fact, many projects are now in progress".—Dr. Cao, BeefLedger researcher

from communication, lead-by-example is also what BeefLedger is trying to do. So far, a few stakeholders participated in the beef blockchain, including producer, abattoir (exporter), and transport company. From the user experience perspective, the blockchain user interface is less developed, which is less easy to use for customers. This is because blockchain is relatively young, and has been mainly developed as a backend system for a long time. Many researchers use the internet as an example and suggest that blockchain will develop into a very capable and accessible system soon (Babich & Hilary, 2018). Consumers will also gradually develop knowledge about blockchain in the near future. The cross-border supply chain also has cultural differences. To understand Chinese consumer behaviours, BeefLedger has conducted research on value perceptions in 2019. The research can provide insights into consumer perceptions and buying behaviours.

Blockchain is still in its infant stage, which means the technology is less mature compared to many other technologies in many aspects. When it comes to building a complex system, which is needed in food supply chains, the weakness of immature technology is obvious. Product information from breeder to the end consumers can be significant, including text, pictures, and videos which require a large storage space and high transaction speed. Data storage and scalability need to be solved along with BeefLedger's growth. BeefLedger is considering and testing various blockchain infrastructures in order to achieve scalability such as Holo chain and BSN (Blockchain Service Network).

Poor networks, which can influence the data collection process and information quality, also pose a challenge. For example, in some cases, 4G network is required for driving the blockchain applications. The limited digitalisation also includes the popularity of hand-filled-in forms and the non-cloud-based information system. This makes the information system less communicable and less efficient with many constraints. Meanwhile, the digital formats of information usually come with CSV or PDF format only that are very old-fashioned. To overcome this challenge, instead of changing everything, BeefLedger took a practical view and developed a data interface that works well for basic CSV and PDF files. Moreover, IoT devices are adopted to capture data to avoid data mistakes. BeefLedger is also working with people to improve local digitalisation progressively where it is necessary and cost-effective.

Building blockchain system requires some specific skills such as coding. Industry experience and knowledge are also necessary. In many cases, skilled software developers can have no industrial experience. Thus, there is a mindset gap between industry practitioners and IT professionals. Technologies need to integrate with existing industry practices and processes. Therefore, a certain level of training is needed for software workers. The training is like a bridge that connects developers and the needs of the real world. In this case, developers can have a better understanding of the operating situation and the real needs, which will also help them in developing some technologies.

Blockchain, as a fundamental technology, has difficulties in accessing information from outside (Babich & Hilary, 2018). It needs sensors to interact with the outside world; therefore, it can combine with the Internet of Things (IoT) devices to achieve better use and efficiency. In this context, BeefLedger has worked with IoT device manufacturers in data collection. For instance, Radio Frequency ID (RFID) is used on the animal ear tag to identify the asset. Other so-called smart tags, which can measure animal temperature, heart rate and blood oxygen to monitor animal health are being developed by many third parties. By scanning the ear tag, the data files can be uploaded to the blockchain automatically. IoT devices can make data-inputting easier and more accurate and mistakes made during manual recording can be maximally eliminated. Risks of products can also be detected as soon as they present. To future enhance data authenticity, a multi-signature system is adopted. Data have to be verified by multiple users before uploading. This is encouraging stakeholders to upload authentic information, as well as preventing data mistakes.

As blockchain is a new technology, regulations, and laws are not yet sufficient. Therefore, global trades need more cautious about each country's policy. How to make information accessible for Chinese consumers is also a challenge. This challenge is unique under China scenario because China firewall blocks information from outside of China. As Chinese customers have trouble accessing information directly from the database which is based on outside of China, BeefLedger has to manually transfer information and store it on a China side server. When Chinese customers scan the QR code that is attached on the beef product, although the information is *from* blockchain, it is not actually *on* blockchain. In other words, there is no automatic link to the blockchain due to China's firewall. Therefore, a cross-chain solution is needed under these circumstances. The restriction from data protection law also brings challenges for BeefLedger in data collection and saving. Therefore, BeefLedger has a dedicated legal team to deal with issues in this area. Another challenge is the intense atmosphere between China and Australia. The turbulent international situation leads to changes in policies. Therefore, BeefLedger needs to maintain a positive attitude and look for more opportunities.

4.6 Case Summary

Overall, BeefLedger is founded as an integrated provenance, blockchain security, and payments platform. It uses blockchain technology to secure provenance information of products and trading confidence of supply chain participants. Warwick noticed the massive issues of food safety and fraud in beef industry, particularly in China. Meanwhile, blockchain technology has gained attention outside of the financial area, and it is believed that it can revolutionise the supply chain by providing transparency and efficiency. After a year and half of internal research, BeefLedger has developed an integrated blockchain-enabled platform that can provide transparency to the beef supply chain to tackle the issues. By now, the POA blockchain ecosystem platform has been completed, and BeefLedger is continuing its platform improvement and commercialisation. This chapter presents the issues within global beef supply chain, and how BeefLedger implemented the platform to solve the issues. The benefits and barriers are also summarised.

References

Babich, V., & Hilary, G. (2018). *Distributed ledgers and operations: What operations management researchers should know about Blockchain technology*. Forthcoming in Manufacturing & Service Operations Management & Georgetown McDonough School of Business Research Paper No. 3131250.

Cao, S, Dulleck, U., Powell, W., Turner-Morris, C., Natanelov, N., & Foth, M. (2020). *Beef ledger Blockchain-credentialed beef exports to China: Early consumer insights*. Technical report, Queensland University of Technology, Australia.

Deloitte. (2020). *Beefing up blockchain: How blockchain can transform the Irish beef supply chain*. Accessed July 05, 2020, from https://www2.deloitte.com/content/dam/Deloitte/de/Documents/Innovation/Beefing-up-Blockchain-Meat-Supply-Chain-Transformation-Deloitte-2018.pdf

European Commission. (2020). *Food fraud: What does it mean?* Food Safety - European Commission. Accessed May 12, 2020, from https://ec.europa.eu/food/safety/food-fraud/what-does-it-mean_en

Grunert, K. G. (2005). Food quality and safety: Consumer perception and demand. *European Review of Agricultural Economics., 32*(3), 369–391.

Hotten, R. (2015). *Volkswagen: The Scandal explained*. BBC News. BBC. Accessed May 12, 2020, from https://www.bbc.co.uk/news/business-34324772

MLA. (2019). *2019 State of the industry report*. MLA Corporate. Accessed July 05, 2020, from https://www.mla.com.au/globalassets/mla-corporate/prices%2D%2Dmarkets/documents/trends%2D%2Danalysis/soti-report/mla-state-of-industry-report-2019.pdf

Pazaitis, A., de Filippi, P., & Kostakis, V. (2017). Blockchain and value systems in the sharing economy: The illustrative case of backfeed. *Technological Forecasting and Social Change, 125*, 105–115.
PwC. (2020). *Fighting food fraud*. PwC. Accessed May 12, 202, from https://www.PwCcn.com/en/industries/food-supply-and-integrity/fighting-food-fraud.html
Saberi, S., Kouhizadeh, M., Sarkis, J., & Shen, L. (2018). Blockchain technology and its relationships to sustainable supply chain management. *International Journal of Production Research, 57*(7), 2117–2135.
Workman, D. (2020). *Top beef exporting countries*. World's Top Exports. Accessed July 05, 2020, from http://www.worldstopexports.com/top-beef-exporting-countries/

Chapter 5
Case Analysis: W Company

This chapter presents the case analysis of W company. It introduces the background of the company; the issues that the company wishes to address; stages of blockchain implementation; the benefits of blockchain; and how the company overcame the blockchain implementation barriers.

5.1 Background Information

This section provides the within-case analysis of W company, including the company background and the issues within the retailing industry in China.

5.1.1 Company Background

Founded in 1962 in America, W company is now the world's largest private employer and retailer, with more than 2.3 million associates and 10,500 stores in 24 countries (Walmart, 2021). Every week, more than 275 million customers are served globally (Walmart, 2021). The revenue in 2019 reached $559 billion, including $10.67 net sales by its China branch (Ecommerce, 2021; Walmart, 2021). W company China branch was opened in 1996, and it ranks as the fourth largest retail brand in China (China Daily, 2019). A total of 434 retail stores were operating in China by the end of January 2021. There are three different formats that are targeting different consumer groups: 396 supercentres, 31 Sam's clubs, and seven Neighbourhood markets. About 96,660 associates are working in about 180 cities nationwide (as of May 2020). A total 7 billion person times were served by the end of 2020 (Walmart China, 2020).

C. Zhang et al., *Blockchain Applications in Food Supply Chain Management*, Contributions to Management Science,
https://doi.org/10.1007/978-3-031-27054-3_5

In 2016, the American branch of W company worked with the IBM Food Trust to use blockchain for food traceability (IBM, 2017). In 2018, all the leafy greens' suppliers were required to register and update information on blockchain-based platforms within a year (Churchill, 2018).

However, the China branch made a different choice. It teamed up with PwC and VeChain in mid-2019 to operate a blockchain platform. Before the project was officially kicked off, the China branch of W company gained interest and discussed it with food industry expert—PwC since 2018. To test and verify the idea, PwC invited its blockchain IT partner VeChain to join in to provide technology service. The official blockchain traceability platform is built based on VeChainThor blockchain, which is a public chain that was developed by VeChain independently.

5.1.2 Background of Main Partners

PricewaterhouseCoopers (PwC), considered as one of the Big Four accounting firms, is a multinational professional services network of firms. With more than 284,000 people in 155 countries and providing services to over 400 of the Fortune 500 companies, PwC ranks as the second largest in its field. It is also the only firm that has a food supply and integrity consulting department in the Big Four. The services include addressing challenges in food safety and quality management, risks assessment, supply chain management, growth strategy, and investments (PwC China, 2021). As early as 2016, PwC developed an interest in blockchain technology and formed a joint business relationship with VeChain in 2018 (VeChain, 2021).

VeChain is a blockchain start-up that was formed in 2015. It has built the Proof of Authority (POA) based public blockchain infrastructure layer—VeChainThor Blockchain. On top of the VeChainThor, VeChain ToolChain was also developed to provide a ready-to-use SaaS platform and RESTful APIs for customising (Cointelegraph and VeChain, 2020; VeChain, 2019). It aims to bring benefits to the supply chain industry with a combination of IoT devices. It has formed partnerships with multiple global companies, such as global auditing company—PwC, Norway's registrar giant—DNV and Chinese insurer—PICC (Wall, 2019). It currently has over 100 staff including more than 50 professionals working in seven offices across Asia, Europe, and North America. VeChain has provided services to many well-known companies in several industries including logistics, retail, and agriculture. It also has won several awards such as the second place for the LVMH innovation Award in Paris (VeChain, 2021).

In this project, the compliance department of W company is the main department that closely works with PwC and VeChain. PwC is playing a role as a project manager, who is responsible for consultancy, business strategy, business process, and stakeholder education. VeChain is mainly responsible for IT development such as coding and platform structuring by requirements.

5.1.3 Retail Sector in China

According to a forecast by a market research company—Emarketer (Cramer-Flood, 2020), China is expected to surpass the USA and became the largest retail market in the world spending 5.072 trillion dollars from 2020 (Fig. 5.1). The huge buying power is due to the rising middle class, rapid growth in e-commerce, and innovative sales formats such as group buying and live broadcasting. However, the physical stores are still accounting for at least 65% of business (Trotter, 2019). One of the top retail trends in China is the advance of using technologies (McKinnon, 2020). Considering merging digital with offline retail, China is 2–3 years ahead of the USA (McKinnon, 2020). Apart from mobile payment, smart shopping carts, self-checkout, and super-fast delivery, consumers can also scan the product's QR code to gain more information about the product, with details including nutritional information, cooking instructions, and consumer reviews (McKinnon, 2020).

5.2 Issues Within China's Retail Supply Chain

This section presents two main issues for W company to address as follows: consumer concerns for food quality and safety, and supplier management efficiency.

Fig. 5.1 The trend of total Retail Sales in China and the USA, 2018–2024 (Source: Cramer-Flood, 2020)

5.2.1 Consumer Concerns

Consumer concerns for products' quality and safety is one of the main issues that W company hopes to address by its blockchain platform. Chinese consumers have strong concerns over food safety and quality after the 2008 Sanlu poisoned infants' milk scandal. According to research by Mintel—a market research company, nearly 10 years after this scandal, 60% of Chinese parents still prefer milk products from Australia or New Zealand, and fewer than a fifth Chinese parents associate Chinese milk brand with a premium image (Desk, 2017). The top four infant formula firms in China that account for 40% of the total market (128 billion yuan) are foreign brands. Moreover, the sales of imported infant milk formula via cross-border e-commerce platform alone were 6 billion yuan in 2014, rising to 18 billion yuan in 2017 (Pak, 2018). The figures suggest that "once bitten, twice shy" is true for Chinese consumers. Subsequent food fraud scandals such as fake beef mixed with pork and chemicals (2013), gutter oil (2014), and fake alcohol (2017) increased consumers' distrust for domestic food products.

The pandemic in 2020 has largely brought consumers' awareness in food quality and safety to a new high level. It has changed the consumption pattern. In a survey of Chinese consumer behaviour, 52% of the participants think that they will shop for healthy food products including vitamins and medicines, and 75% of the Chinese consumers are going to adopt a healthier lifestyle (McKinnon, 2020). According to the China food safety development report in 2019, over 50% of the food incidents in China were caused by human factors, such as microbial contamination, excess use of food addictive, quality failure to meet standards, pesticide residue failure to meet standards, and heavy metal contamination (CFDA, 2020). In this case, larger food producers and retailers are trusted more by consumers compared to smaller food retailers (Kendall et al., 2019). This also means that as a world-leading food brand, W company has more responsibility to provide quality and safety products, which will enhance consumer confidence as well as help strengthen its market competitiveness to differentiate itself from competitors.

The concept of food traceability is not new in the market. Over a decade ago, traceable food could be found on the market with QR codes. However, a report by Bjfood (2014) showed that consumers were not fully satisfied about the food traceability at that time. The QR code was usually made by the food company itself. The information on the traceability platform was either too little, or not what consumers wanted to know. Mostly, traceability was only a way of publicity, which was useless during food recall. Another popular traceability method was to reply on a third-party traceability platform. Information about food products from raw material, production, logistics, and sales on the platform could be checked by both consumers and regulators (Bjfood, 2014). Although using a professional third party sounds reliable, both the credibility and efficiency of the third party are questionable (Grunert, 2005).

5.2.2 Supplier Management

The China branch of W company works with approximately 7000 suppliers in China (Walmart China, 2020). The large number of suppliers brings difficulties for efficient supplier management including quality control and supplier audit. To manage suppliers, W company not only had supplier standards to regulate suppliers, but also formed the ethics and compliance department specifically for investigations and assessments since 2011 (Walmart China, 2020). The department is deigned to control risks within 14 areas including food safety, product safety, licences and permits, and responsible sourcing. To ensure food safety, the compliance team is not only responsible for educating suppliers about the regulation and laws, and checking their certifications and permits, but also have the rights to audit and inspect suppliers' factories at any time. Apart from its internal audit, W company has also employed a third-party auditor to simplify the auditing process from 2017. Third parties can review documents and visit supplier sites. Suppliers with higher risks are required to submit audit reports more frequently.

According to the Co-founder of VeChain—Mr. Zhang, *"early on, W company had to set up a complete supplier management standard, including access mechanism, regular audit and test. But there are unreliable factors such as cutting corners or quality issues. It was difficult for W company to obtain supplier performance data. Because W company only gets to know if a supplier is good or not when there is an incident. The cost to maintain a dedicated team to conduct regular inspections and spot checks on the suppliers is also huge"*. The investments in compliance are massive; there are over 2000 associates globally, and more than $269 million has been spent in compliance enhancements from 2016 to 2018 alone (Oh et al., 2019; Walmart, 2020). Although millions of dollars and thousands of workers were invested to prevent potential risks, it is still hard to achieve 100% security. It is impossible for compliance teams to inspect every detail 24 h a day, 7 days a week, particularly when the labour force is limited but the number of suppliers is huge. Capital, human resources, and time can be wasted during the process. For example, suppliers may receive inspections according to the product volume rather than the type of products and suppliers' past behaviours.

Moreover, there are still possibilities for dishonest behaviours such as data tampering and corruption. The supplier performance data are not sufficient nor reliable, because suppliers used to upload information manually and leave the gap for data tampering. According to the senior manager at PwC food & beverage retail—Mr. Feng, *"the product origins used to be manually entered, and can be changed in the background anytime. This kind of data changing is happening quite often in the food industry. For example, we all know that pears from Xinjiang are delicious, the suppliers can change the pear origins to Xinjiang. This is the weakness of traditional traceability method"*. Moreover, corruption may also exist both in internal and third-party audits. Documents, permits, and certifications can be expired or fake. Once food safety issues happen, it can lead to disputes between suppliers and W company because the original data may be changed at any stages. Data authenticity could be the key to responsibility partition. To conclude, the challenge for W company China branch is how to achieve a better efficiency in human resource utilisation and allocation while managing supplier issues.

5.3 Blockchain Implementation

This section presents the blockchain implementation process from a practice-based view (Table 5.1 and Fig. 5.2), including the preparation, what has been done, and the plan for future.

5.3.1 Preparation

Two Proofs of Concept (POC) of blockchain food traceability were announced by W company in 2016: tracing the origin of mangos and tracking the pork in China stores (Fig. 5.3). In the pork tracking trial, W company teamed up with IBM and Tsinghua University and tracked pork from the factory to the retailer. The successful trial largely increased the confidence for W company to use blockchain in all its stores. However, rather than just providing technology, if the system builder has enough industry experience to give adequate consulting is one of the key considerations for the China branch W company. Mr. Feng mentioned, *"it is not purely about technology, the food industry is complicated. So, during the implementation, if you do not understand this industry, it is going to be very difficult. PwC is here because we have a dedicated food team, we are the only one in the Big Four that has a consulting team specialising in food industry. Combining our industry and consulting knowledge, and we have a close business relationship partner—VeChain, all those factors made W company decide to work with us"*. Therefore, when PwC joined the project, VeChain was also invited to provide technology solutions.

The China branch of W company did not try to implement blockchain until 2018, when Mr. Feng met the senior managers of W company China branch in an industry conference. "*I have always been in the food industry, and I have known some of their managers for nearly ten years. We firstly talked about these concepts (blockchain) in an industry conference. Turns out that they were very interested. Therefore, we kept discussing it and trying to find the commercial values of it and proposing and verifying business plans for over half a year. Then, we started to set up the project*", said Mr. Feng. PwC is a desirable partner for W company. It not only has a dedicated food team that specialises in the food industry, but also has a close partnership with VeChain—a blockchain start-up company that was invested in by PwC. Earlier in 2018, VeChain already provided several blockchain solutions for tracing Italian wine brands and formed a partnership with global assurance company—DNV. The rich experience that VeChain has can also provide a great help to W company in platform building.

Table 5.1 An overview of blockchain implementation stages

Stage	Period	Stakeholders	Resources (who owns what)	Actions	Outcomes and expected outcomes
Set-up stage	Mid 2018–May 2019	W company PwC VeChain	1. Past blockchain experience (POCs from 2016)—W company 2. Research reports—PwC 3. Workshops between key business partners for brainstorming—W company, PwC, VeChain	1. Invite technology provider—VeChain, Food industry consultant—PwC, and identify piloting Tier 1 suppliers 2.Test various business plans (W company, PwC, VeChain)	1. Successfully reached an agreement among stakeholders 2. Contract signed in May 2019 3. Tested and selected 23 product lines
Stage 1	May 2019–May 2020	W company Piloting Tier 1 suppliers PwC VeChain	1. Supplier conference—W company, PwC, VeChain 2. Sales report—W company 3. Transaction volume analysis report—VeChain	1. Selecting suppliers and products (W company) 2. Connecting and motivating suppliers (W company and PwC) 3. Educating suppliers (PwC) 4. Building the blockchain system (VeChain)	1. Platform was successfully built with 100 products live on chain 2. Invited more than 20 Tier 1 suppliers to participate
Stage 2	June 2020–present	W company All Tier 1 suppliers Piloting upstream Suppliers PwC VeChain	1. Supplier feedback 2. Sales report—W company 3. Transaction volume analysis report—VeChain	1. Connecting and motivating more suppliers including some sub-suppliers (W company and PwC) 2. Educating suppliers (PwC) 3. Optimising the system (VeChain) 4. Changing targeting consumers (W Company and PwC)	1. All Tier 1 suppliers and all products will be traceable by blockchain 2. Some sub-suppliers will be invited to join the platform 3. System functions are going to be enriched and the efficiency is going to be optimised 4. The targeting group is transferring from supercentres to Sam's club
Future plan	In a near future	W company All Tier 1 suppliers Piloting	1. Supplier feedback 2. Sales report—W company	1. Inviting more upstream sub-suppliers 2. Enabling	1. The scale of suppliers is expected to expand

(continued)

Table 5.1 (continued)

Stage	Period	Stakeholders	Resources (who owns what)	Actions	Outcomes and expected outcomes
		upstream Suppliers PwC VeChain	3. Transaction volume analysis report—VeChain 4. Legal opinion about data privacy—VeChain	consumer feedback module 3. Enabling smart contract for instant payment 4. Launching a micro-insurance project with PICC (People's Insurance Company of China)	2. W company hopes to lawfully collect some consumer feedback and consumer information 3. Smart contract is expected to streamline the financial process between suppliers and W company 4. The food products that are traceable on blockchain platform and have DNV certificate are expected to have micro-insurance to enhance the food security.

Eventually, the official contract was signed in May 2019, which marks the start of the project. Twenty-three product lines were first tested and listed on the platform, and the products range expanded gradually in the following years. On 25 June in the same year, W company displayed its performance during the China Food Safety Publicity Week and announced that the W company's blockchain traceability system was officially live.

5.3.2 Implementation

5.3.2.1 Stage One

The first stage lasted a whole year from May 2019 to May 2020. The project started in Shenzhen, with 23 products in W company's supercentres being tested and listed on the blockchain platform. By the end of 2019, more than a 100 products in over 10 categories were traceable on the platform. By the end of 2020, the blockchain-traceable meat accounted for 50% of the packed meat sales, traceable leafy greens

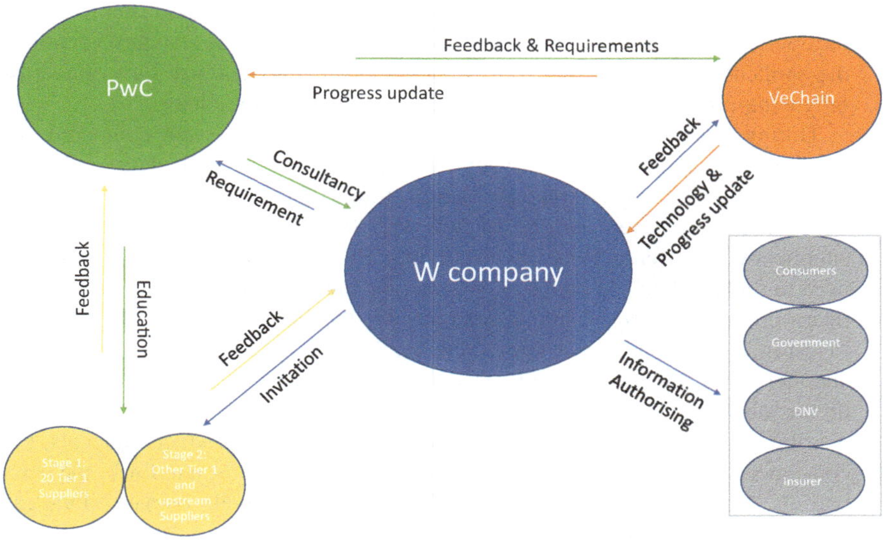

Fig. 5.2 W company's blockchain implementation diagram (Source: the author)

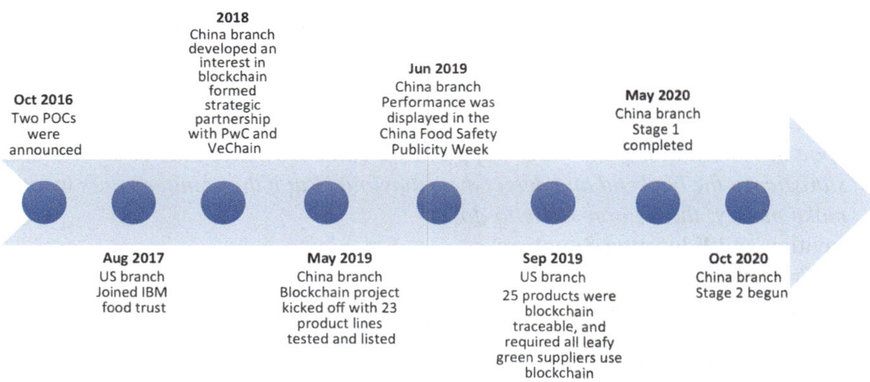

Fig. 5.3 Timeline for blockchain project (Source: Hyperledger, 2021)

accounted for 40% of the packed vegetables, and traceable seafood accounted for 12.5%.

1. Structuring the System

   At the technology level, Stage 1 is mainly about platform building, which is mainly overseen by VeChain. Mr. Cheng, the project manager in VeChain, explained that *"in fact, we are more about constructing the entire system. For example, if W company has certain requirements on the functions of the system, we will fulfil the needs. Secondly, we need to ensure the system runs stable. So, the operation needs to be guaranteed by us, that's what we are focusing on".*

Moreover, Mr. Zhang also said that in order to make it easier and faster to deploy the application, VeChain has made some improvements to the standard version of the application. *"This is our goal. We hope our clients can use the application fully within a week from the start of the project"*, said Mr. Zhang.

During platform building, the data collection methods were compared and considered carefully. W company had considered several data collection methods, such as radio frequency identification (RFID) and QR code. QR code was eventually chosen because of low costs, which is affordable for different suppliers with different financial constraints. By using QR code, suppliers do not need to replace their old facilities or invest in a new one. For large suppliers with many production lines, barcode scanners are the only equipment they need to invest in. The scanners only cost a few hundred Chinese yuan each. For smaller suppliers with few production lines, a smart phone with dedicated APP can also do the job. Suppliers can upload activity records by scanning the QR code, while distribution centre can also scan the code to confirm the products' arrivals. For consumer user experience, QR code to web page is also chosen to lower the entry barriers. Compared to a dedicated APP, any consumer with a smart phone can scan and check product information.

To against potential QR code counterfeit, W company chose to allocate every product a unique QR code instead of one QR code for a batch. Therefore, every product has a unique and random QR code, which can largely eliminate potential QR code-faking behaviours. According to Mr. Feng, "*for example, there are 1000 products in a batch, then we have 1000 different QR codes randomly allocated to each product. Therefore, if people want to fake the code, they will have to either buy the whole batch which may contain 1000 products, or copy one code in 1000 times. If the latter happens, we can easily check the code-scanning statistics in the backend and detect the behaviour. But if the counterfeiters want to make money, that is non-sense to do*".

2. Inviting and Educating Suppliers

From the stakeholder perspective, in stage 1, W company focused on how to persuade Tier 1 suppliers to get on board. There are over 7000 suppliers, thus it is important for W company to think carefully about what products should be traced on blockchain and who should be invited first. According to Mr. Feng, the criteria for selecting first-stage products are fresh products, hot-selling products, products that consumers care about the most, and the products with suppliers who have the higher passion to participate. For example, beef and seafood are selected in the first stage. Moreover, at the first stage, only Tier 1 suppliers were invited. This means that consumers can trace products up to the product factory side. The information can include location, logistics, and production batch. However, more information beyond Tier 1 suppliers such as raw material suppliers was not available at this stage. Once suppliers agree to participate in the project, PwC can provide guidance and sufficient training to selected suppliers about how to use the blockchain system.

5.3.2.2 Stage Two

Stage 2 was expected to start from June 2020. However, the outbreak of coronavirus impacted the project and postponed it for a few months until October 2020. In this stage, a few things were changed and updated from stage 1. Mr. Feng summarised, "last year (2019), we were only selling products in W company's supercentres, but now, Sam's club is also joining in. The functions of the system are enriched and optimised in each stage as well. More suppliers are joining us, and we are continuing to provide them on-site training".

1. Changing Target Groups from Supercentres to Sam's Club

   Although W company started the blockchain system from the supercentres, it now puts more efforts into the membership-only high-end store—Sam's club. The target consumers of Sam's club are the middle- and high-income families. Compared to supercentres, customers in Sam's club are less sensitive to prices but have more expectations on product quality. They are more open to the latest technologies such as blockchain traceability system. Therefore, from mid-2020, Sam's club launched its own blockchain traceability platform, which is based on VeChain ToolChain. Many products from Sam's club's own brands such as Member's mark are on the chain, including imported wine, seafood, and meat products. Consumers can track the information (such as information about logistics, processing, quality certificates, supplier information) up to Tier 1 suppliers by scanning the QR code. According to a report that was published by VeChain (2020),

   > In the future, more products will be included on the platform, providing consumers with transparent and checkable "information receipts", helping the brand continue to consolidate the market image of high-end food retailer and expand the industry influence.

   To attract more customers, W company has also provided some guidance and publicity about the blockchain platform in the store.
2. Optimising the System

   From the technical side, stage 2 is about system optimisation. Since the project has been going on for a year, W company has gained more experience and received feedback from suppliers which were further passed on to VeChain. According to Mr. Cheng, *"in this project, system building is not very difficult, the difficulty is how can you persuade suppliers to join in. Some companies might just want to have a blockchain-based internal information system, and they can simply require all departments to use it. However, for W company, the platform participants are not directly under its control. Moreover, the precondition for many suppliers to adopt the system is that the system should negatively impact their production efficiency and logistics. That is why we say that the key things we need to consider during system implementation are the system efficiency and the sustainability of the operation"*.

   In the second stage, interface design is also emphasised. For example, the web page background colour for seafood is blue, and the colour tends to green while scanning vegetables. The improvement in interface design can bring convenience to consumers and encourage them to scan the code to obtain information. Just as Mr. Zhang said, *"we are trying to make consumers feel like they are just using normal software, I believe this is the friendliest way for consumers"*.

3. Inviting More Suppliers

   More suppliers are invited to use the blockchain system, including some upstream sub-suppliers. Compared to the first stage, stage 2 allows a longer supply chain traceability, which means that products can be traced back to Tier 2 or even lower-tier suppliers. For instance, beef products can now be traced to the feedlot, and rice products can be traced to the plantation farm. Therefore, more details about the products are gradually added on the chain. Moreover, the scale of suppliers also expands from national to international. Imported goods can also be traced back to their origins.

5.3.3 Future Plan

This section lists a few plans that W company is considering for the future, including inviting more upstream sub-suppliers, enabling consumer feedback, and activating smart contract.

5.3.3.1 Inviting More Upstream Sub-suppliers

By the end of stage 2, all Tier 1 suppliers and part of the upstream suppliers should be part of the blockchain project. In the future, W company hopes to involve more suppliers and list more products on the platform. Those are the things that W company is continuing working on since the beginning, although the outbreak of coronavirus has postponed the project. Fortunately, according to Mr. Zhang, the negative impact of COVID-19 was very limited to the project process, which is back to normal at present. In other words, W company will constantly extend the supply chain traceability both vertically and horizontally. As Mr. Cheng mentioned, *"we will discuss with W company gradually about what to do next. The only certain thing in the future is that W company has plans to involve more suppliers and products. The other things are constantly adjusting to achieve continuous improvement"*.

5.3.3.2 Enabling the Module of Customer Feedback

Due to the restrictions of data protection law, detailed consumer information cannot be collected at present. Therefore, how to collect consumer feedback without breaking laws and regulations is still a challenge. According to Mr. Zhang, *"the module of consumer feedback was considered in the earliest blueprint. But we had to hold on to it for the future because of the user privacy laws. We do not know the user experience from the consumer side, the only feedback we received is from the suppliers and W company*". VeChain is seeking advice from law firms and trying to balance between W company's requirements and data protection law. Currently, only some rough and general consumer data can be collected, such as when the consumer scanned the code. Moreover, the future uncertainties on the regulation are also the challenges that VeChain needs to cope with constantly.

5.3.3.3 Using Smart Contract

Smart contract may be used in the future to automate and streamline the financial process between suppliers and W company. Instead of paper contract, smart contract allows business partners to complete transactions and send payment automatically based on the pre-agreed contract. According to Mr. Feng, *"for example, in the past, we usually had to conduct a multi-party verification to confirm the receiving of goods, and then payment can be initiated. The whole process is long. But if they start to use smart contract, as soon as suppliers ship the products and W company's distribution centres receive the products in the right amount and appropriate condition, the payment can be initiated. This is because the data of the products are recorded on the blockchain platform permanently. So, people have trust and no need to confirm so many things again. In other words, once the smart contract is set up, we can synchronise the completion of the delivery and billing in both financial and legal sense"*.

5.3.3.4 Micro-Insurance Project

There is also an insurance company—PICC (People's Insurance Company of China) that has joined the partnership with DNVL and VeChain. It is planning to launch a micro-insurance project for the products that are both traceable on blockchain and approved by DNV. As Mr. Zhang introduced, *"by adding a little bit of money on top of the food price, it is equivalent to every consumer buying an insurance at the same time as they buy food. If there are food safety issues, such as food poisoning, the insurance company can compensate based on the insurance policy. Although there might be unexpected incidents occasionally, the insurance company can still make a profit from it, because both blockchain and DNV have provided the credibility"*.

5.3.4 Stakeholder Engagement

To encourage suppliers to participate in this project, W company promised them better places to list their products. For example, a notable area that is particularly for blockchain-traceable products is labelled the "trust-free" zone. According to Mr. Zhang, *"with such an 'trust-free' area, suddenly, consumers bought out all of these products. Consumers think that scanning and seeing information is a fun thing to do, and they believe what W company suggests. With these special arrangements, more suppliers are willing to join in. It starts to become a positive cycle and brings in more suppliers"*. W company also invited suppliers, PwC and VeChain to a supplier conference for discussion and education of blockchain. Despite some complex technology issues, PwC is in charge of supplier education. "*Suppliers need to use our APP and need to log in to the background for information uploading. Our team needs to teach suppliers how to use the system and how to upload information, online or on-site"*, Mr. Feng added.

VeChain, as the technology provider, however, does not have direct contact with suppliers. Before the project kicked off, VeChain was only in charge of platform building. Once the platform was launched, VeChain can optimise the system based on suppliers' feedback that was collected by W company and PwC. Mr. Cheng said, "*we only need to persuade W company to launch the project. As soon as they approve the project, they communicate with their suppliers, especially the suppliers who have good relationships with them. They persuade them from the perspectives of food safety and marketing*".

Apart from incentives and education, another key motivator for suppliers to take part is the efficiency of the application. For suppliers, the system they are willing to try is the system that has no negative impact on their production efficiency. According to Mr. Cheng, "*why we say that the key to operate and apply blockchain application depends on the construction of the entire system, and if the system has a minimum impact on suppliers' production efficiency? If the impact is little, they are usually happy to cooperate with W company to do things. This is also an opportunity to market their products at the same time*". Moreover, the fact that VeChainThor Blockchain is a public blockchain is also important for suppliers to know. Compared to private chains, public blockchain means that no one is dominant on the chain, even though W company is initiating and leading the blockchain project. It provides a strong credibility of the platform that suppliers will not lose their power or need to worry about data alteration.

Apart from PwC, other third parties such as government, DNV, and insurance companies are neither on the chain nor directly involved in the project. PwC is the only directly involved third party to check the data authenticity and inspect factories. However, other third parties can also benefit from the blockchain platform (Fig. 5.4). For regulators, they can access information when necessary. Sufficient traceability information can efficiently help them to deal with food scandals. Mr. Cheng explained, "*the government actually pays attention to it. At the national level, companies are encouraged to do such things (blockchain-enabled traceability), because it is meaningful for people's livelihood"*.

DNV, a global quality assurance and risk management company, has formed a partnership with VeChain (Morris, 2018). By combining blockchain and IoT devices, DNV will be able to make sure of the data authenticity and audit data remotely. The certification statement from DNV not only attracts more consumers as it provides credibility, but also encourages suppliers to actively participate in the project. The platform will also benefit DNV as auditing on behalf of the China

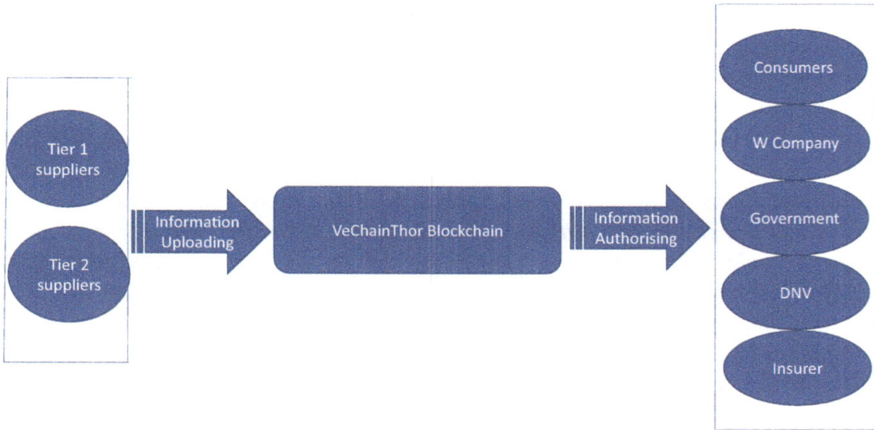

Fig. 5.4 Information flow (Source: the author)

branch of W company is a new business. Different from its existing industry coverage such as maritime, energy and electrification, food and beverages are a relatively new opportunity for DNV.

5.4 Blockchain Benefits

This section presents the benefits that were brought by blockchain, such as solving the abovementioned issues, improving food transparency and traceability, and increasing the supplier management efficiency. There are also extra benefits from the marketing perspective, such as attracting more consumers, increasing industry influence, and improving brand reputation. Mr. Cheng summarised, *"the purpose of the project has two dimensions, one is from the food safety dimension which also has marketing benefits, another one is the internal dimension which is supplier management"*.

5.4.1 Transparency and Traceability

The first benefit that blockchain brought to W company is the transparency. Transparency is always a challenge in the food supply chain. Different products need different levels of transparency. For fresh products, consumers care more about the transportation information. For expensive products, the product origins are the key information. Consumers who shop for products such as sustainable sourced products, organic food, and halal food all have strong demands on transparency.

Blockchain can be used to provide more transparency, including the stories behind the product, logistics information, quality certificates, etc. The transparency will give consumers more buying confidence in this case. Under the same price, consumers will tend to buy products with the traceability label (Galvez et al., 2018).

"Blockchain has backed up food traceability", according to VeChain (Zuifen, 2020). Sufficient information stored on blockchain can significantly accelerate food recall and remove the problematic products. Mr. Cheng said: *"Instead of the traditional paper records, which is very inefficient with very little information on it, they (regulators) can see some detailed records on Blockchain if there is a quality accident"*. This is particularly helpful during the pandemic to trace contaminated products and reduce further impact. Therefore, W company can use the blockchain system to recall faulty products speedily; meanwhile, consumers can also scan the QR code to know if the products are safe to eat.

5.4.2 Supplier Management

5.4.2.1 Precise Supplier Management

Another benefit of blockchain is to improve its supplier management efficiency. As mentioned above, W company not only formed an ethics and compliance group for supplier inspection, but also started to use third parties for auditing. However, the efficiency of people assignments can be very low, while the labour costs are expensive. By using blockchain, W company can make better use of its human resources and precisely manage the suppliers at different levels depending on the supplier performance data. Factories with high potential risks can be required to submit audit reports and receive inspections more frequently. Mr. Cheng explained, *"if they have 1 people to do factory inspection, for example, now they can rearrange tasks and re-allocate resources more effectively. For instance, they can reduce the inspection for lower-risk suppliers from quarterly to semi-annually and increase the site visiting frequency of higher risk factories. Therefore, the utilisation efficiency for personnel resources is greatly improved"*.

Moreover, the adoption of blockchain also turns brand trust or trust in the relationship to technology trust. This transformation is particularly useful and necessary for new cooperation between two parties. The technology, therefore, can impose restrictions on company behaviours and encourage legitimate cooperation. Because the blockchain platform can also help W company to find the root cause by ensuring sufficient supplier data, it can also help W company to manage risks from preventing potential unlawful actions.

5.4.2.2 Liability Defining

Apart from classified and precise supplier management, blockchain can also help to define liability and solve potential supplier disputes. One of the useful features of blockchain is data immutability. Once data are uploaded on the chain, they will be kept permanently with a time stamp. Data correction or withdrawal will be recorded. This means that if food incidents happen, blockchain can not only help to find the problem, but also can be used as evidence to claim responsibility and solve disputes between W company and suppliers. *"Before, suppliers give information to W company and let W company display the information to consumers. However, if something goes wrong, W company can simply change or delete the information. By blockchain, as the data are immutable, both parties can have more trust in each other. If there is a dispute, they can match the product data with the data on the chain to define liability"*, explained by Mr. Cheng.

The technology itself cannot guarantee lawful and honest behaviour; however, the data on blockchain will not be changed or erased and is like a "warning sign" that can remind business partners to behave legitimately. Like Mr. Zhang said, *"blockchain is only an underlying technology, the technology itself cannot do anything. However, the immutability and security features make the data that blockchain carries have a certain value"*.

5.4.3 Marketing

Apart from solving problems, blockchain can also attract more consumers and enhance brand image. From scanning the code, consumers can obtain more information about the products. The immutability feature of blockchain can also guarantee the information authenticity. In the market filled with food scandals, the more transparency in the food supply chain, the more buying confidence consumers will have. Qualify certificates from DNV and micro-insurance can further strengthen the consumer confidence.

For W company, the blockchain-enabled traceability system will greatly promote the brand image of selling quality products. As a leading company to use the blockchain traceability platform in food industry, the technology will provide a strong competitive advantage, particularly during the coronavirus pandemic when people care more about food safety.

5.5 Implementation Barriers

This section presents how W company overcame the barriers during implementation from the following perspectives: intra-organisational barriers, inter-organisational barriers, systems-related barriers (Table 5.2).

Before the project was officially launched, W company has gained experience from running trials. Once it decided to fully adopt blockchain, W company invited PwC—experts in food industry and VeChain—experts in blockchain technology to jointly build the blockchain platform. Over half a year's discussion on the company demands, market trends, and business plans, the blockchain project was officially kicked off in mid-2019. W company hopes to enhance food safety and improve supplier management efficiency.

A few years ago, a dedicated department was formed for ethics and compliance for auditing needs. Third-party audits are also applied to increase credibility and simplify the audit process. W company showed a strong determination since then. Not only were test runs conducted, but also went through a half-year long discussion before officially signing the platform-building contract. Millions of Chinese yuan were spent on platform building, supplier education, and publicity since then. To ease the financial pressure of suppliers, W company has covered most of the costs. It also works with PwC to provide sufficient training to suppliers.

W company had conducted two POC projects in 2016. The successful experiences had offered W company deep insights about the value of blockchain. China branch had invited two partners, one with strong food industry experience and the other with knowledge of blockchain technology, to jointly build the platform. Both experienced and professional business partners greatly helped W company while launching the project.

In most cases, adopting innovation requires large financial investment. In this case, W company has invested millions of Chinese yuan in this project in platform building, supplier education, and publicity. However, the investment is affordable for W company as it was the world's largest company by revenue in 2019. Moreover, the following maintenance fee is also relatively low. Moreover, the increased efficiency and enhanced food safety management are worth the investment.

From the first stage, W company had provided incentives and education to attract its Tier 1 suppliers to use blockchain. For example, suppliers can list their blockchain-traceable products in "trust-free" area and receive quality certificate from DNV, which can provide products more credibility. They can also participate in a supplier conference to discuss the blockchain system with W company. Once they agree to participate, PwC provided sufficient guidance about how to use the platform. Considering most of the suppliers are localised, there are no barriers in communication and cultural difference. Many suppliers worry about the system efficiency and data privacy. To solve the concerns, VeChain is constantly optimising and adjusting the system according to the supplier's feedback. Moreover, suppliers are not required to upload sensitive information apart from performance data and

Table 5.2 Barriers from four perspectives

Four perspectives	Barriers categories	Barriers in practice	Approaches and solutions	Supporting quotations
Intra-organisational barriers: the barriers in this category are from the internal of the organisation	Company demand ambiguity	Hard to find the company needs	W company invited PwC and VeChain and discussed and tested various business plans over half a year to find the best fit before the project was officially launched. Meanwhile, the strategy and platform design stay constantly updated to adapt to the market needs.	"Rather than a technical consultation, PwC is doing food safety consultation. We know what the system should look like to meet food safety requirements in the market. Regardless of the food safety internal control, or external promotion, I know what to do for the best. For example, what kind of information that suppliers should upload to meet internal control. PwC is giving suggestions to W company based on the market trends".—Mr. Feng, senior manager at PwC food & beverage retail.
	Challenge in organisational culture, management, and leadership	Hard to keep a strong determination	W company showed a strong determination from two test runs in 2016, and half a year's discussion since 2018. Millions of Chinese yuan were invested since then to build the platform, educate suppliers, and publicity. A dedicated department was formed for ethics and compliance for auditing needs a few years	"I have always been in the food industry, and I have known some of their managers for nearly 10 years. We first talked about these concepts (blockchain) in an industry conference. Turns out that they were very interested. Therefore, we kept discussing it and trying to find the commercial values of it and proposing

(continued)

Table 5.2 (continued)

Four perspectives	Barriers categories	Barriers in practice	Approaches and solutions	Supporting quotations
			ago. Third-party audits are also applied to increase credibility and simplify the audit process.	and verifying business plans for over half a year. Then, we started to set up the project".—Mr. Feng, senior manager at PwC food & beverage retail.
	Limited number of use cases	Lack of experience to try an emerging technology	Two POC projects that were successfully completed in 2016 have offered W company valuable experience. W company also invited two partners with strong industry experience and knowledge.	"W company, American branch and Chinese branch had already tried out some POC with IBM on blockchain from 2016 to 2017".—Mr. Feng, senior manager at PwC food & beverage retail.
	Financial constraint	Massive investment	W company was the world's largest company by revenue in 2019. Although the initial investment is about millions of Chinese Yuan the following maintenance fee is affordable. The increased efficiency and enhanced food safety management are worth the investment.	"The fee is composed by the cost of continuous system development, the fee of consulting solutions on this project, and supplier education fee. For supplier coaching fee, the more suppliers, the more fees will be applied. The three parts of the fees are the main part of W company's initial costs".—Mr. Feng, senior manager at PwC food and beverage retail.
Inter-organisational barriers: the barriers are between	Supplier engagement	Lack of incentives to implement blockchain	W company provides incentives to encourage suppliers include	"With such an 'trust-free' area, suddenly, consumers bought out

(continued)

Table 5.2 (continued)

Four perspectives	Barriers categories	Barriers in practice	Approaches and solutions	Supporting quotations
the supply chain stakeholders			better locations to blockchain-traceable products, such as "trust-free" area and DNV quality certificates.	all of these products. Consumers think that scanning and seeing information is a fun thing to do, and they believe what W company suggests. With these special arrangements, more suppliers are willing to join in. It starts to become a positive cycle and brings in more suppliers".—Mr. Zhang, Co-founder of VeChain.
		Lack of sufficient knowledge of blockchain	A supplier conference was held to discuss blockchain. To engage more suppliers, sufficient education and guidance about blockchain are also provided by PwC.	"Suppliers need to use our APP, and need to log in to the backend for information uploading. Our team needs to teach suppliers how to use the system and how to upload information, both online and on-site".—Mr. Feng, senior manager at PwC food & beverage retail.
		Cultural and geographical difference	W company started the project with its Tier 1 suppliers who have close relationship with W company, plus 95% of the products are sourced locally (Walmart China, 2020). This means	"We only need to persuade W company to launch the project. As soon as they approve the project, they communicate with their suppliers, particularly the suppliers who have good

(continued)

Table 5.2 (continued)

Four perspectives	Barriers categories	Barriers in practice	Approaches and solutions	Supporting quotations
			that there is neither geographical difference nor cultural difference between suppliers and W company.	relationship with them. They persuade them from the perspectives of food safety and marketing".—Mr. Cheng, the project manager in VeChain
		Efficiency concern	Feedback from suppliers are collected and reported to VeChain for system optimisation.	"The biggest problem we encountered during implementation is how to improve the efficiency. In this area, for example the information input, they have repetitive tasks every day. We had to consider how to reduce the repetition and the workload. That's why we say that the key to operate and apply blockchain application depends on the construction of the entire system, and if the system has a minimum impact on suppliers' production efficiency. If you need a dedicated staff to spend a long time on the system, plus many errors can be easily made during data input which leads to rework, if this is the case, suppliers will have a lot of concerns about using the system.

(continued)

Table 5.2 (continued)

Four perspectives	Barriers categories	Barriers in practice	Approaches and solutions	Supporting quotations
				If the impact is little, they are usually happy to cooperate with W company to do things. This is also an opportunity to market their products at the same time".—Mr. Cheng, the project manager in VeChain
		Data privacy concern	The blockchain platform is built on VeChainThor, which is a public chain. Suppliers and W company share equal power on the chain; neither party is dominant.	"We use the public chain, so in a practical sense, W company do not own all the data, nor have a chance to change the data. So, you can make suppliers believe that there is no problem to upload data. The credibility of the public chain is much higher than other kinds. Moreover, a data encryption process is involved; this means that suppliers upload data in an encrypted form, not in plain text. If there is no special arrangement, W company cannot see all the data".—Mr. Zhang, Co-founder of VeChain.
			Suppliers are not required to upload sensitive information, only performance data and	"Suppliers will not give sensitive information to W company. Only the information

(continued)

Table 5.2 (continued)

Four perspectives	Barriers categories	Barriers in practice	Approaches and solutions	Supporting quotations
			production data are required.	for quality control, and information they are willing to show consumers are on the chain. For example, inspection reports, annual audit reports and qualification certificates, these non-sensitive types of information".—Mr. Cheng, the project manager in VeChain
	Financial pressure for suppliers	Concerns to share costs and update facilities	Suppliers do not share the costs of platform building, education costs, and publicity	"Because W company want to make a faster move, so they decided to cover all the costs, including supplier training and consulting costs".—Mr. Feng, senior manager at PwC food & beverage retail.
			Suppliers only need to buy affordable facilities (one code scanner per production line/smart phone with an APP).	"We have done a very accurate cost calculation. The suppliers only need to buy a code scanner per production line. The scanner can as cheap as from 200 Chinese yuan. For smaller suppliers, if the production is not much, the smart phone with an APP can also do the job".—Mr. Feng, senior manager at PwC food & beverage retail.

(continued)

Table 5.2 (continued)

Four perspectives	Barriers categories	Barriers in practice	Approaches and solutions	Supporting quotations
			To reduce costs, W company chose to use barcode over RFID.	"With QR code, it costs a few cents per label. You still need to consider the costs. For example, five cents are 1% of the gross profits for a box of vegetables that cost five Yuan (1 Yuan = 100 cents). If we use RFID, the costs are at least ten times of QR codes".—Mr. Feng, senior manager at PwC food & beverage retail.
	Lack of consumer participation	Lack of consumer awareness	To better serve the consumers, Sam's club has even developed its own dedicated blockchain platform. In addition, there is also publicity in the store to provide guidance of blockchain system to consumers.	"For Sam's club, its target group is mid-to-high-end consumers. They are not very sensitive to the price, but are more sensitive to the product quality, such as country of origin. So, there is some special publicity in the store to tell consumers that the products are traceable on the platform. The products have traceable labels".—Mr. Cheng, the project manager in VeChain
			VeChain also considered how to be more friendly to consumers. Rather than downloading a dedicated APP, consumers can	"We are trying to make consumers feel like they are just using normal software, I believe this is the friendliest way for

(continued)

Table 5.2 (continued)

Four perspectives	Barriers categories	Barriers in practice	Approaches and solutions	Supporting quotations
			simply scan the QR code that is linked to the website. The background colour of the information page also changes depending on the products.	consumers".—Mr. Zhang, Co-founder of VeChain.
Systems-related barriers: the barriers relate to the technology.	Technology immaturity	Scalability issue	To avoid data crowd, VeChain has combined online and offline data. Data are wrapped up, encrypted, and stored offline on W company's own server. Only the hash, which corresponds to the data, is stored online. Other extra information such as pictures are directly generated from the Internet.	"We did not store all the information on chain because it will slow it down. While the consumer is scanning the QR code, there is always a few seconds verification process with the blockchain background. If the code is incorrect, the information will fail to display. Once the verification succeed, consumers not only can see the encrypted product information, but also can also see a blockchain transaction hash (TXID), which can be verified on VeChain's public blockchain".—Mr. Feng, senior manager at PwC food & beverage retail. "When consumers finally see the products, the information has already been uploaded a long

(continued)

Table 5.2 (continued)

Four perspectives	Barriers categories	Barriers in practice	Approaches and solutions	Supporting quotations
				time ago. This is not like financial payment that requires a short time to complete. It is not necessary for this business scenario. In terms of the business transaction volumes, even larger volumes, this platform can still handle the transactions very well".—Mr. Cheng, the project manager in VeChain
		Products with mixed ingredients are hard to track	Processed products, unlike fresh products, usually have to go through multiple steps with mixed ingredients. For processed products, such as cooking oil, it is impossible to trace the whole supply chain. W company only needs to provide information including product batch, quality report, production data, etc. Those product details are sufficient for both consumers and regulators.	"The question is what kind of information do consumers want? For products that they pay more attention to the product origin, such as Manuka honey, we will have to chase it up the whole supply chain. But for industrial products, we only need information such as production batch, inspection report, production data. These kinds of information are basically enough. The other information can be controlled by its internal traceability system. There is no need to track all of the ingredients".—

(continued)

Table 5.2 (continued)

Four perspectives	Barriers categories	Barriers in practice	Approaches and solutions	Supporting quotations
				Mr. Feng, senior manager at PwC food & beverage retail.
	Immutability with erroneous data	Date inputting mistakes by accident	To prevent such mistakes, W company and PwC have regular audits on factories to check the products and data authenticity. The immutable record can also encourage suppliers to pay more attention on data inputting.	"It is more effective to formulate the blacklist system, so once the data has been uploaded, you cannot change it, you will have to be responsible for it. We (PwC) as a third party to check data authenticity, W company might also do it themselves".—Mr. Feng, senior manager at PwC food and beverage retail.
			If there are mistakes already made that cannot be changed, the emergency action is to hide the wrong data from consumers (whilst both right and wrong data are on the chain).	"Of course, to prevent some mistakes that maybe complained by consumers, typos such as grams instead of kilograms, the one thing we can do is to hide the wrong information. Instead, we link the traceability system to the official website".—Mr. Feng, senior manager at PwC food & beverage retail.
External barriers: the barriers from external environment.	Lack of government involvement and encouragement	Government attitude impacts	China government shows a supportive attitude and encourages food companies to	"Government actually pays attention to it. At the national level, companies are

(continued)

Table 5.2 (continued)

Four perspectives	Barriers categories	Barriers in practice	Approaches and solutions	Supporting quotations
			adopt new technologies to improve food traceability. During the pandemic, China's government is paying even more attention to tackle food safety and traceability issues by applying technologies such as blockchain.	encouraged to do such things (blockchain-enabled traceability), because it is meaningful for people's livelihood".—Mr. Cheng, the project manager in VeChain
	Issues relate to policies and regulations	Restriction from data protection law	In this project, VeChain seeks advice from law firms to align with policy requirements. W company also hopes to be able to collect consumer feedback in the future.	"The module of consumer feedback was considered in the earliest blueprint. But we had to hold on to the module for the future because of user privacy laws. We do not know the user experience from the consumer side, the only feedback we receive is from the suppliers and W company".—Mr. Zhang, Co-founder of VeChain.
		Relevant policy deficiency	Data protection laws such as GDPR (General Data Protection Regulation) and general cybersecurity law can apply to all technologies in all areas, including blockchain. In this case, there are governmental policies in China, such as	"We need our clients to propose a boundary about what kind of the information they want to collect. We want to make them happy without violating data protection law".—Mr. Zhang, Co-founder of VeChain.

(continued)

Table 5.2 (continued)

Four perspectives	Barriers categories	Barriers in practice	Approaches and solutions	Supporting quotations
			Cybersecurity law of the People's Republic of China in 2016 (Lawinfochina, 2016). The law is expected to guarantee cybersecurity and information-based development. It should apply to any use of a network in China. Blockchain as an information system is also included in this law.	

product data. The blockchain platform is operating on the public chain, which also largely enhances the data security.

Suppliers are not required to share the system development costs nor education costs. W company has covered most of the investment. The only cost for suppliers is the cheap barcode scanners or a smart phone APP to scan code and upload data. The blockchain system is designed to be easily used, which will not impact the production efficiency. The blockchain-traceable products will be presented in a better area and attract more consumers. In other words, suppliers not only can adopt innovation with low cost, but can also attract more consumers and increase sales.

To engage more consumers, W company has provided publicity and guidance about the blockchain platform in the store and decided to use a QR code instead of a dedicated APP. The QR code can bring more convenience to consumers and increase their willingness to scan the code. Consumers can reach to a website with product information simply by scanning the QR code. The other functions also added to make it more user friendly, for example, the background colour of the website varies depending on the products.

One of the technological issues of blockchain is the limited storage. To overcome potential storage issues, only the corresponding hash will be uploaded onto the chain, other information will be encrypted and saved on the off-chain iCloud server. Consumers scanning the QR code will need to go through a verification process. Other extra information such as pictures are directly generated from the Internet. Another issue specifically in the food industry is that it is hard to achieve full traceability for products with mixed ingredients. These products, therefore, are only required to be traceable to the factory side. More detailed information from upstream suppliers is not necessary.

For information mistakes, whether on purpose or by accident, VeChain can temporarily hide the information from consumers to avoid complaints or misleading. On the one hand, if suppliers upload fake data, PwC and W company will send teams to conduct factory inspections and to check data. On the other hand, the immutability feature of blockchain is meant to encourage suppliers to behave honestly and be more careful on data inputting.

External stakeholders are also being supportive in the project. For regulators, once there is a food issue, sufficient information on the blockchain platform can accelerate the food recall process, as well as find out who is responsible. Therefore, the Chinese government has showed strong interests about exploring blockchain potentials in the food industry. China's President Xi spoke highly of blockchain on a political collective study and suggested that China should seize the industrial innovation opportunity to accelerate the blockchain development (Foxley, 2019). The unique features of blockchain (decentralisation, immutability, and smart contract) are particularly useful during the pandemic for food recall.

Although there is not yet a dedicated policy for blockchain technology, there are new issued general data protection laws in China that can still apply to this case. The law protects the stakeholders' rights but brings difficulties for VeChain to collect consumer information. Therefore, one of the plans in the future for W company is to have an active consumer feedback module within the legal range. In order to do so, VeChain is actively seeking legal advice from the legal department of W company and PwC.

5.6 Case Summary

Overall, W company China branch formed a strategic partnership with food industry expert—PwC and blockchain technology professional—VeChain to improve food traceability and supplier management. During the blockchain implementation, the coronavirus pandemic in 2020 has warned Chinese consumers again about food safety. Consumers have shown higher interests in knowing more about the food they eat. Therefore, the blockchain project has been carried out as planned under this situation. The project is currently in its second stage. This chapter discussed the issues that W company hopes to address, different blockchain implementation stages, the benefits that blockchain brought, and how W company overcame the barriers from the four perspectives: intra-organisational, inter-organisational, systems related, and external perspective.

References

Bjfood. (2014). *Domestic food traceability is in vain? How can we go from virtual to real*. Accessed 07.02.2021, from http://www.bjfood.net/MobileShowNews/NewsDetail/6708

CFDA. (2020) *<China food safety development report (2019)> proposes that China's food safety faces five types of risk challenges*. Accessed 03.02.2021, from http://www.cfda.com.cn/newsdetail.aspx?id=128469

China Daily. (2019). Top 10 supermarket brands in China in 2019. *China Daily*. Accessed on: 03.02.2021, from https://www.chinadaily.com.cn/a/202007/08/WS5f04f9c2a310834817257d8d_1.html

Churchill, F. (2018). *Walmart tells leafy greens suppliers to use blockchain*. Accessed 22.05.2019, from https://www.cips.org/en-GB/supply-management/news/2018/october/walmart-rolls-out-blockchain-to-all-leafy-greens-suppliers/

Cointelegraph and VeChain. (2020). *When blockchain meets IoT: Ensuring food safety in the 2020s*. Accessed 02.02. 2021, from https://s3.cointelegraph.com/storage/uploads/view/4612f245e3cf8bc227084dff049f4016.pdf

Cramer-Flood, E. (2020). *China Ecommerce 2020*. Emarketer. Accessed 02.02. 2021, from https://www.emarketer.com/content/china-ecommerce-2020

Desk, N. (2017). *Organic products "driving growth" in Chinese infant formula – study*. Accessed 22/01/2021, from https://www.foodbev.com/news/organic-products-driving-growth-in-chinese-infant-formula-study/

Ecommerce. (2021) *Is there Walmart in China?* Accessed 22/05/2021, from https://adebayothevoice.com/qa/is-there-walmart-in-china.html

Foxley, W. (2019). *President Xi says China should "seize opportunity" to adopt blockchain*. Coindesk. Accessed 22/01/2021, from https://www.coindesk.com/president-xi-says-china-should-seize-opportunity-to-adopt-blockchain

Galvez, J. F., Meijuto, J. C., & Simal-Gandara, J. (2018). Future challenges on the use of blockchain for food traceability analysis. *Trends in Analytical Chemistry, 107*, 222–232.

Grunert, K. G. (2005). Food quality and safety: Consumer perception and demand. *European Review of Agricultural Economics, 32*(3), 369–391.

Hyperledger. (2021). *Case study: How Walmart bought unprecedented transparency to the food supply chain with Hyperledger Fabric*. Accessed 01/03/2021, from https://www.hyperledger.org/learn/publications/walmart-case-study

IBM. (2017). *Walmart, JD.com, IBM and Tsinghua University Launch a Blockchain food safety Alliance in China*. IBM. Accessed on: 14.11.2018, from https://www-03.ibm.com/press/us/en/pressrelease/53487.wss

Kendall, H., Kuznesof, S., Dean, M., Chan, M., Clark, B., Home, R., Stolz, H., Zhong, Q., Liu, C., Brereton, P., & Frewer, L. (2019). Chinese consumer's attitudes, perceptions and behavioural responses towards food fraud. *Food Control, 95*, 339–351.

Lawinfochina. (2016). Cybersecurity law of People's Republic of China. *China's Leader in Online Legal Research*. Accessed 22/01/2021, from http://www.lawinfochina.com/display.aspx?id=22826&lib=law&encodingname=big5

McKinnon, T. (2020). *The top 8 retail trends in China to watch*. Indigo9Digital. Accessed 03/02/2021, from https://www.indigo9digital.com/blog/top-shopping-trends-from-china

Morris, N. (2018). Blockchain VeChain partners with Chinese insurer PICC. *Ledger Insights*. Accessed 22/01/2021, from https://www.ledgerinsights.com/blockchain-VeChain-china-insurance-picc-dnv/

Oh, A. Y. K, Mendelsohn, M., Karp, B. S. (2019). *Walmart's failure to maintain a sufficient anti-corruption compliance program*. Harvard Law School Forum on Corporate Governance. Accessed 25/01/2021, from https://corpgov.law.harvard.edu/2019/07/16/walmarts-failure-to-maintain-a-sufficient-anti-corruption-compliance-program/

Pak, J. (2018) *Foreign infant milk formula still highly converted in China 10 years after the melamine scandal.* Accessed 22/01/2021, from https://www.marketplace.org/2018/10/24/foreign-infant-milk-formula-still-highly-coveted-china-10-years-after-melamine/

PwC China. (2021). *Food supply and integrity services.* Accessed 22/01/2021, from https://www.PwCcn.com/en/industries/food-supply-and-integrity.html

Trotter, C. (2019). What's happening with retail in China?. *Insider Trends.* Accessed 04/02/2021, from https://www.insider-trends.com/whats-happening-with-retail-in-china/

VeChain. (2019). *VeChain Whitepaper 2.0.* Accessed 03.02.2021, from http://www.VeChain.org/qfy-content/uploads/2020/01/VeChainWhitepaper_2.0_en.pdf

VeChain. (2020) *The "Last Kilometre" of blockchain commercial landing: Combination of standardized products and customed tools.* Accessed 22/01/2021, from https://mp.weixin.qq.com/s/iUH25S84QJ2yT6o6jhYzlA

VeChain. (2021). *About us.* Accessed 22/01/2021, from https://www.VeChain.com

Wall, J. (2019). *VeChain will be hard to stop when PwC, Deloitte, BMW, and China's financial insurance giant are involved – A sleeping giant?.* Accessed 04/02/2021, from https://www.investinblockchain.com/VeChain-will-be-hard-to-stop-when-PwC-deloitte-bmw-and-chinas-financial-insurance-giant-are-involved-a-sleeping-giant/

Walmart. (2020). *Global Ethics & Compliance.* Accessed 22/01/2021, from https://corporate.walmart.com/our-story/global-ethics-compliance

Walmart. (2021). *About us.* Accessed 22/05/2021, from https://corporate.walmart.com/our-story

Walmart China. (2020). *Walmart China Factsheet.* Accessed on: 22/01/2021, from http://www.wal-martchina.com/english/walmart/index.htm#china

Zuifen. (2020). *Walmart's Chinese subsidiary cooperates with VeChain to track food.* Accessed 22/01/2021, from https://www.zuifen.com/block/662242.html

Chapter 6
Case Analysis: FairChain

This chapter presents a within-case analysis of FairChain. The content includes background of the company and the industry; the issues that FairChain hopes to address by applying blockchain; the actual implementation stages and the plan for future; what kind of benefits blockchain can bring to FairChain; and what the challenges are that FairChain overcame.

6.1 Background Information

This section introduces in what situation FairChain Foundation was built and the current European coffee market.

6.1.1 Company Background

FairChain Foundation and Moyee coffee were founded in 2013 by Guido van Staveren van Dijk in Netherlands. The idea of FairChain popped into his head when he read some research about the value distribution inequality in the coffee value chain (Fig. 6.1). "*The attitude of I do not care about if the person who grows my coffee can make a proper living out of that, or if they can get to school or get some proper health care, et cetera. That's what really makes us sick*", said by Mr. R, the chief technology officer of FairChain Foundation. To build an equal value chain and to help the local coffee farmers, van Staveren van Dijk in Netherlands gathered up a few private investors and applied for government funding in the meantime (less than 50% of the total investment) to build a FairChain Foundation, and to sell its commercial product—Moyee coffee since 2013. Moyee coffee shop was opened in Dublin in 2016, the business has now expanded to the whole of Europe.

C. Zhang et al., *Blockchain Applications in Food Supply Chain Management*, Contributions to Management Science,
https://doi.org/10.1007/978-3-031-27054-3_6

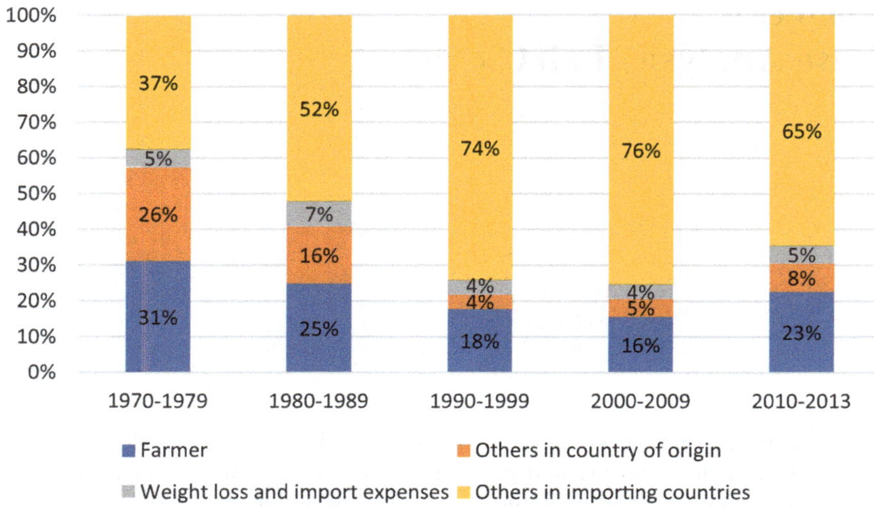

Fig. 6.1 Coffee value distribution before 2013 (Source: Samper & Quinones-Ruiz, 2017)

Moyee coffee is the world's first FairChain coffee brand, which is aiming to restore the balance between coffee producer and consumers. Going a step further than fair trade, FairChain and Moyee coffee move the value-added processes such as roasting to the country of origins—mainly in Ethiopia. It started with financing a roastery in Addis Ababa (2013) Ethiopia to help farmers earn added value (FairChain Foundation, 2014). The goal is to keep at least 50% of the profit in the country of origins; in other words, the local farmers, roaster and local communities are accounting for 50% of the chain profit. However, balancing the value chain sounds like a dream story that most people will not believe the validity of. Therefore, FairChain has to prove itself to the world and the authenticity of the story. That was the reason why blockchain technology was adopted in this project in 2018. The blockchain platform was jointly built by FairChain and KrypC on Hyperledger. It was officially launched from 2018.

To cover the costs in this Moyee coffee project, FairChain foundation not only provides a blockchain solution for Moyee coffee, but also provides software solutions to other companies in the market. Mr. R has also invested FairChain and claimed that "*we are still working to reach our break-even point. So, we are extending the number of clients, and hope to be fully self-supporting in 2022–2023. We are on the right track to get there. It is not only just spending all this money, but it is spending our own money, so we are really focusing on making this business a success*".

6.1.2 Global Coffee Supply Chain

As one of the most frequent traded commodities and one of the most common drinks globally, over 50 countries are participating in coffee production by growing, processing, roasting, etc. (Kucukcay, 2020). However, the coffee supply chain is complex and broken. There are usually 13 stages within a coffee supply chain from farmers to consumers (Fig. 6.2). It is noticeable that many middlemen (up to 20 middlemen and 6 months) operate between farmers and consumers (Ashoka, 2014). This is because most coffee growers are smallholders (1–2 ha), who have no direct access to the market and very little bargaining power. Therefore, farmers have to reply on the middlemen to link multiple stakeholders and bring their coffee beans to the market. Among all the intermediaries, roasters are earning most of the profit in the value chain (Ashoka, 2014).

Moreover, to make a cup of coffee, there are 15 steps to complete (Coffee Bean Corral, 2021) (Fig. 6.3). The length of the whole production process can last 4–7 years as growing coffee cherry requires specific warm climate (Kucukcay, 2020).

Fig. 6.2 Typical coffee supply chain (Source: Field, 2014)

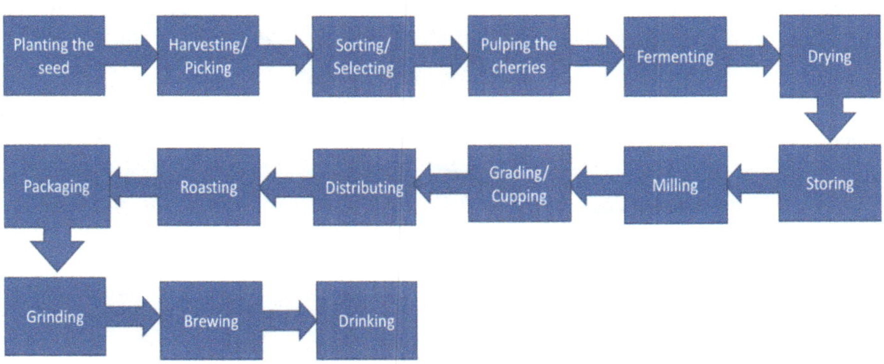

Fig. 6.3 Coffee making process (Source: Coffee Bean Corral, 2021)

The environmental condition can strongly impact coffee tree growing. Mostly, coffee only harvested once a year, and the harvest season requires huge labour investment to pick the coffee fruits. However, after all the time, efforts, and labour invested in coffee growing, farmers can only earn as little as 0.4% of the coffee price (Bruce-Lockhart & Terazono, 2019; UNCTAD, 2020).

6.1.3 European Coffee Industry

Coffee is one of the most frequently traded commodities in the world, and an estimated 1.6 billon cups of coffee are brewed every day. The coffee industry has reached $102.15 billion in 2019 and is expect to be valued at $155.64 billion by 2026, by an average increasing annual rate of 5.5% by 2025 (Businesswire, 2020; CBI, 2020). Europe is a large coffee market; in 2019, it consumed 3,356,000 tonnes of coffee (average over 5 kg per person per year) and accounted for about one-third (34%) of the global coffee consumption (CBI, 2020) (Fig. 6.4).

The Netherlands, a medium-sized coffee market in Europe, accounts for 5.6% (averagely 8.4 kg per person per year) of the consumption in Europe (CBI, 2020). However, as a leading coffee market of certificated coffee, the Fairtrade certificated coffee only accounts for about 3% of the coffee market share, and only half of the coffee has a sustainable certification in the Netherlands (CBI, 2019). Fairtrade is a certification scheme that indicates coffee producers are paid at the Fairtrade minimum price ($1.4 per pound or $1.7 per pound for organic). It exists to guarantee the global minimum price for coffee.

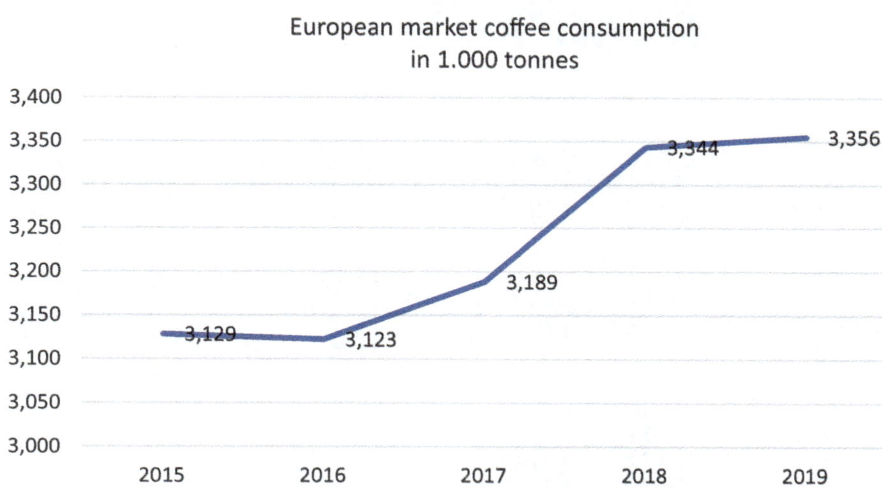

Fig. 6.4 European coffee market consumption from 2015 to 2019 (Source: CBI, 2020)

Moreover, the total European market imported over 3.7 million tonnes of green coffee beans in 2019; the largest importers were Germany (34%), Italy (19%), Belgium (9.3%), and Spain (8.4%) (CBI, 2020). Europe also has a large coffee roasting industry that is dominated by a few large companies such as Nestle, Jacobs Douwe Egberts, and Lavazze (CBI, 2020). Thus, Europe is also the largest roasted and ground coffee exporter (84% of the total roasted coffee export volume in 2018) in the world (CBI, 2020). The Netherlands accounts for 6% of the green bean importers and 10% of the roasted coffee exporters (CBI, 2019, 2020).

6.2 Issues Within Coffee Supply Chain

This section discusses the main issues within the coffee industry that FairChain hopes to address by blockchain. The issues include value inequity due to the out-of-balance value chain and unsustainable sourcing.

6.2.1 Value Inequality

Globally 125 million people depend on coffee for their livelihoods, and 25 million smallholders produce 70–80% of the world's coffee (Fairtrade, 2021). However, the average income of coffee farmers has stayed the same in the past 20 years. Most of the smallholder coffee farmers live under the international poverty line (44%) which is $3.20 per day or even extreme poverty line (22%) which is $1.9 per day (Rushton, 2019). Farmers only earn 0.4% of the price per cup of the coffee (Bruce-Lockhart & Terazono, 2019; UNCTAD, 2020) (Fig. 6.5). Moreover, nearly 61% of the coffee producers are selling coffee at a very low price which is even lower than the production cost (Fairtrade, 2019).

In this case, Ethiopia, is considered as the birthplace of coffee, as suppliers of Arabica coffee and organic certified coffee, which accounted for 2.5% of the total European coffee imports in 2019 (CBI, 2020). However, the poverty rate was 70.50% in 2015, and 32.6% of the population were under the extreme poverty line (Knoema, 2021). In 2017, only about half of the population (51.77%) could read or write (Macrotrends, 2021).

One of the reasons that local coffee producers earn so little is that most value-added activities have been moved out of the country of origins. For example, according to CBI (2019), most roasters are based in the Netherlands after importing green beans. Coffee producers can only gain a minimum salary for tree planting and cherry pick-up while most value-added processes (roasting and packing) cannot benefit the local area. Figures show that the market prices for coffee are between 6.29 € per kg for lower-end to 39 € per kg of higher-end; however, the export prices of green coffee only account for about 5%–25% of the selling price (CBI, 2019).

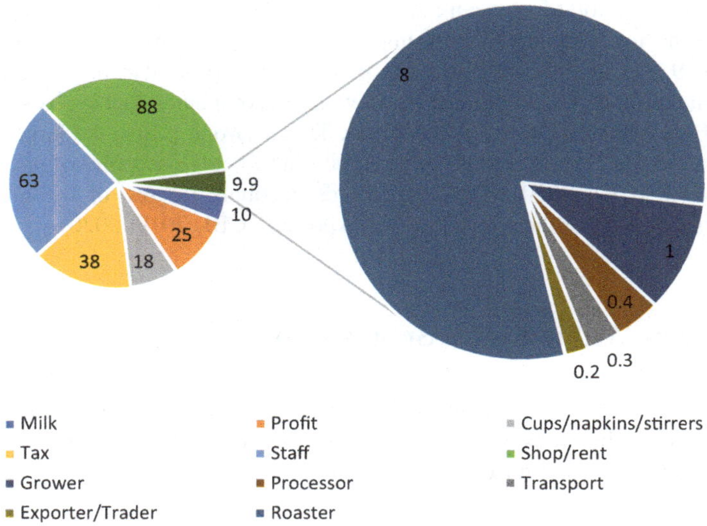

Fig. 6.5 Value distribution in one cup of coffee in pence (£2.5/cup) (Source: Bruce-Lockhart & Terazono, 2019)

6.2.2 Unsustainable Sourcing

Farmers are living in poverty while producing coffee to satisfy the increasing world coffee market. However, the low income not only brings farmers difficulties in normal life, but also lead to a vicious cycle between poverty and unsustainability. There are three principles of sustainability: profit (economy), people (sociality), and planet (environment).

First, from the *economic* perspective, poverty means that farmers cannot afford new equipment to improve productivity or conduct value-added activities. Therefore, in most cases, farmers can only sell the coffee cherries, which has very little added value. The earnings for the farmers are one-time-only depending on how many harvests there are each year. Moreover, weather change also has strong impact on coffee planting. Drought or excessive humidity can reduce the production significantly, which leads farmers to a worse financial situation. "There *is an inconsistency in the harvest. Last year (2018) they had a lot of rain and the whole harvest basically failed. So these are real-world problems",* said Mr. A, the FairChain Architect.

Second, from the *social* perspective, the low income can directly affect farmers' living conditions. The living income is defined as the decent amount of money that can cover standard expenses on living, housing, food, healthcare, education, transportation, and unexpected issues. Therefore, farmers who are earning less than the living income can be seriously threatened. Poverty also means that farmers cannot afford proper education. Therefore, deficiency in literacy usually comes with low living income, which can make farmers' situation even worse. The low literacy, in this project, has also become one of the practical challenges during blockchain implementation for FairChain. Moreover, lack of education can also lead to child labour and forced labour, which are also common problems in the coffee industry.

Third, from the *environmental* perspective, poverty means that farmers want (need) to grow more plants but cannot afford to change plots or properly fertilise their land. The consequence of the intensive cultivation can lead to unproductive land. Barren lands not only lead to unproductivity, which means less income for farmers, but also can cause serious environmental issues. To produce coffee in a sustainable way, it is important to invest in new equipment and replant new plots (Fairtrade, 2019). The development in facilities and planting areas can improve the production and the earnings and reduce the environmental impact.

6.3 Blockchain Implementation

This section introduces different stages of blockchain implementation by FairChain, including preparation, actual implementation, how to engage stakeholders, and future planning (Table 6.1, Fig. 6.6). According to Mr. R, 20/80 is the current percentage of the value chain, while only 20% of the added value stay at the origins. FairChain is aiming to restore the coffee value chain to a 50/50 balance. Different from many companies that use blockchain for food traceability and safety, FairChain is using blockchain for value redistribution by balancing the coffee value chain. According to Mr. R, "*we don't use a blockchain as a reconciliation thing for air freight, for waters, et cetera. We don't do that. For us, it's a pure provenance of the product value distribution. That's, for us, the interesting part*".

6.3.1 Preparation

Before the blockchain platform was officially applied in 2018, FairChain carried out a few semi-structured interviews with local coffee producers in Limu, Ethiopia in 2017 (FairChain Foundation, 2019a). The interviews aimed to collect data about local household level to understand their lives. The results show that 350 families who participated (average seven people per family) grow coffee on small plots, and most of them live under the poverty line (less than $2 a day) or the extreme poverty line (less than $1.9 per day) (FairChain Foundation, 2019a, b). Most coffee is grown in semi-forest conditions, with a 1–3 ha average farm size and 150 kg green beans yield per hectare. Moreover, there is no farmer organisation or proper training to farmers. There is no direct access to the market which leaves farmers vulnerable to price bargaining. The supply chain inefficiency and too many middlemen also make farmers lose about 40% of the export value. The lack of access to the washing station makes farmers miss out on value-added activities.

Table 6.1 Overview of the implementation stages

Stages	Technology diffusion processes	Period	Stakeholders	Actions	Resources	Outcomes and expected outcomes
Preparation	Knowledge/ awareness Persuasion Decision	2017–2018	1. FairChain 2. Moyee coffee 3. Producers 4. NGOs 5. Bext360	1. Conduct a semi-structured interview (2017) and created "FairChain farming living income roadmap" 2. Piloting with Bext360 (end of 2017–early 2018) 3. Digitalising the "first mile" (2017)	1. Government funding 2. Help from Non-Governmental Organization (NGOs) and local communities (local team and local universities)	1. Three hundred and fifty families grow coffee on small plots, most of them live under the poverty line or extreme poverty line. 2. A pilot was conducted to trace real-time payment 3. From 2017, information on farmer level is on the chain
Implementation	Implementation	2018 to present	1. FairChain 2. Moyee coffee 3. Coffee producers 4. NGOs 5. KrypC 6. CropIn Technology	1. Continuing "first mile digitalisation", add more information (2018), enabling SmartFarm APP with CropIn (2018) for data collection 2. Enable tokenisation by KrypC (2018) 3. Provided one-to-one financial training to farmers (2018) 4.Conducting a focusing group interview on digital literacy rate (2019)	1. Experience from the pilot 2. Help from NGOs and local communities (local team and local universities) 3. Feedback from stakeholders	1. From 2018 onwards, payment slips are on chain, data collection method was digitalised by using SmartFarm APP, and the interview results suggest that the local area suffers from high illiteracy and low smart phone possession. 2. Supply chain started to be tokenised, and more farmers accepted banking system. In 2020, 400 new farmers started to use bank services and could access to digital wallet

Future	Confirmation/ continuation	In a near future	1. FairChain 2. Moyee coffee 3. Coffee producers 4. NGOs 5. KrypC 6. CropIn Technology	1. Contacting PwC for a continuous auditing 2. More sustainable solutions will be operated 3. Expanding the products scale 4. Moving to a public chain	1. Experience gained from previous stages 2. Feedback from stakeholders 3. Help from NGOs and local communities (local team and local universities)	1. FairChain expects to replace annual auditing report to a continuous auditing on blockchain. 2. FairChain will step further on sustainability, not only on social impact, but also on environmental and economic perspective, such as solar power and recycle waste for more income 3. More products will be on chain as well, including Otherbar chocolates, and honey project is in planning. 4. FairChain is trying to build a platform on public chain, which will allow the public to access and will benefit more on the transparency.

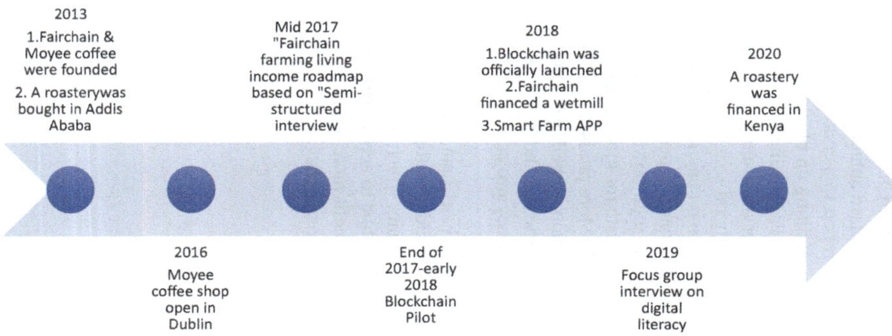

Fig. 6.6 Project timeline (Source: the author)

Based on the results of the interviews and relevant research, FairChain created a "FairChain farming living income roadmap". It is a framework to find out the opportunities to help farmers improve their living conditions and living incomes. The framework was put into practice in three ways from 2017: "Outgrowers' program" for agricultural training, blockchain-empowered value chain, and farmer collaboration on large-scale plantations.

From 2017 to 2018, an "Outgrowers' program" was conducted to provide training to the farmers about how to improve yield and quality by plant rejuvenation and soil fertility. The programme is supported by coffee experts from the local university and a university from the Netherlands. More specifically, the training includes compost making and application, and rehabilitation of coffee trees (FairChain Foundation, 2019a, b). The productivity is expected to double after putting the training into practice and renovating the farms.

Later in 2017, FairChain teamed up with blockchain consultancy company—Bext360 for a pilot blockchain project to trace coffee supply chain. The pilot combined artificial intelligence (AI) and smart contract to enable automatic coffee bean quality evaluation and payment issuing to farmers straight after cherry collecting. Coffee can be traced from the washing station in Ethiopia to the retail store in Europe in a totally transparent approach.

FairChain also started to run a "the first mile digitalisation" from 2017. As Mr. R explained, "*the first mile means that the farmer has a face. It's not just a farmer, but we know all our farmers by name, we have their faces and their profiles. We know what they're producing, how big their families are. And now we're capturing data that will contribute to our long-term ambition that to bring them to a living income level*". In other words, farmers are becoming visible in supply chain with details on the chain. The information can include farmer details, their land size, and tree numbers.

6.3.2 Implementation

1. The "first mile digitalisation" was functional since 2017, and more information was digitised from 2018, including the proof of payments and payment slips. Costs started to be traceable since then. Another improvement is changing the way data are collected. A SmartFarm app was adopted later in 2018 to digitalise data collection process (CropIn Technology, 2020; FairChain Foundation, 2018). Instead of using pen and paper, the information can be collected and sent to the blockchain platform accurately and quickly by a smart phone with SmartFarm APP. Due to the instability of the local network connection, FairChain has combined online and offline methods. Data can be collected offline if necessary, but will be synchronised online as soon as there is network.
   During the "first mile digitalisation", FairChain spent 4 weeks conducting 60 interviews with 30 farmers in mid-2019 (Dekker et al., 2019). The interviews aimed to investigate the local literacy rate and smart phone possession rate. The results showed that there are high illiteracy rates and bad phone networks (Fig. 6.7). The interviews were divided into four specific groups: young female, young male, old female, and old male. Only young males have both high literacy rate (92%) and high phone possession rate (77%). Other groups either have low literacy rate, or most of them do not own a smart phone.
   As Mr. R said, "*Luckily the new sort of generation and the sons and daughters of the farmers are on their way. And that's also a big opportunity because they know about smart phones, they know about technology, they know what technology can do for them. So they are very eager to join us because they see there's a huge benefit of running things on blockchain. Those challenges become an opportunity when the next generation comes into the blockchain*". Therefore,

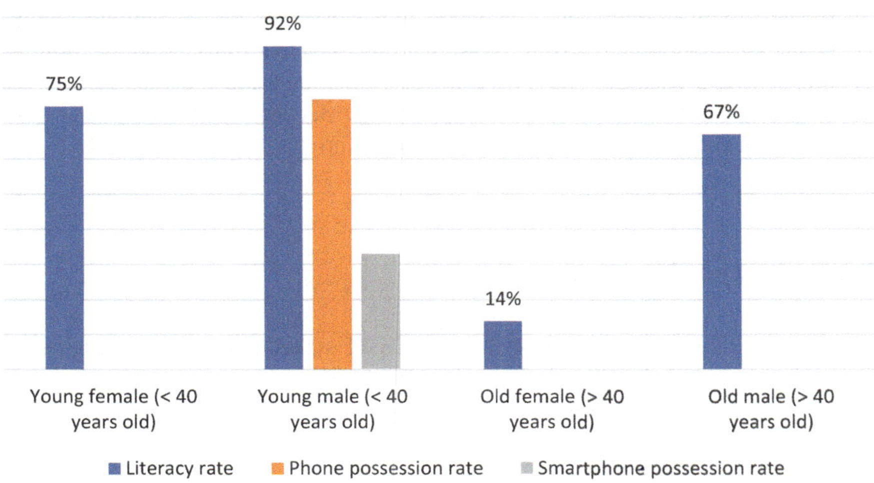

Fig. 6.7 The interview results of four focusing groups (Source: Dekker et al., 2019)

new generations play an important role in the project. They are more open minded and eager to try new things to change their lives. Moreover, the results of the interviews provided FairChain with deeper insights of local household lives, which helped to improve the interface design and the user experience. For example, a listening function can be used if the farmer cannot read (Dekker et al., 2019).

2. Blockchain technology expert—KrypC joined FairChain project in 2018. According to Mr. R, "*I sourced it via a company called KripC, and they have extensive knowledge on Hyperledger and now to build blockchains on that. And together we design blockchain. So, we do the service design together and they build the intellectual properties for FairChain*". The KrypC blockchain platform can create cryptotokens that correspond to a certain value of the commodity (FairChain Foundation, 2019c). The value will increase when the beans go through the coffee supply chain (FairChain Foundation, 2019c). Moreover, consumers will also receive a digital loyalty token after purchasing. The loyalty token can be used for personal use (discount for next order), or it can be shared with farmers for tree planting (Stokes & Reilly, 2019).

   One of the main challenges to enabling blockchain tokenisation is that farmers have a strong distrust for banks (FairChain Foundation, 2020). Before FairChain can tokenise the process, it is necessary to encourage farmers to open bank accounts and accept digital wallets. In order to do so, FairChain has provided one-to-one financial training to farmers and invited bank staff to speak with farmers directly. After the training, farmers not only can gain trust in the banking system but can also learn how to access the digital wallet by their Unique ID (FairChain Foundation, 2020). In 2020, 400 new farmers opened bank accounts and unique ID and can access their digital wallet.
3. As FairChain is using to prove equal value distribution, the majority of stakeholders involved in the value chain are on the chain. So far, the activities on the chain are planting/harvesting, washing process (wet mill), grading, hulling, roasting, and packing, warehouse in the Netherland and logistics in between (Fig. 6.8). Consumers as the last mile of the coffee supply chain are not on the chain but can enhance the impact and tip the farmers.

6.3.3 Stakeholder Engagement

This section introduces how FairChain engaged stakeholders; from farmers, to consumers, to external stakeholders such as NGOs (Fig. 6.9).

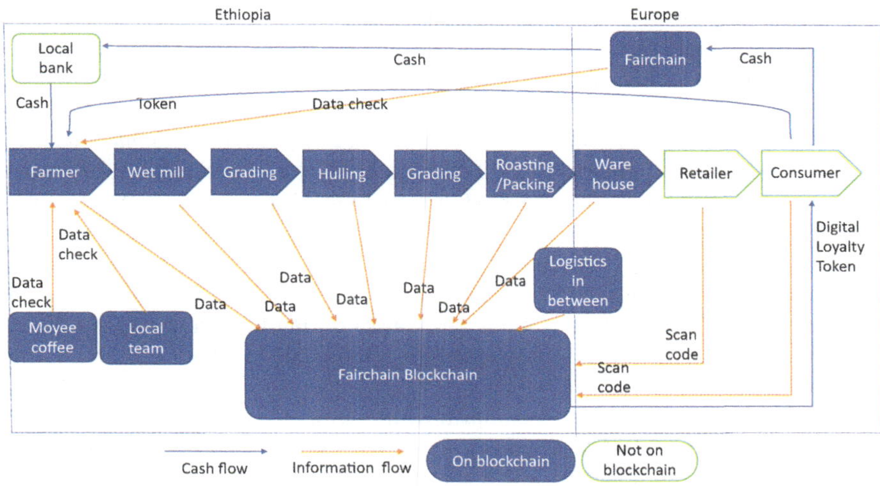

Fig. 6.8 FairChain blockchain framework (Source: the author)

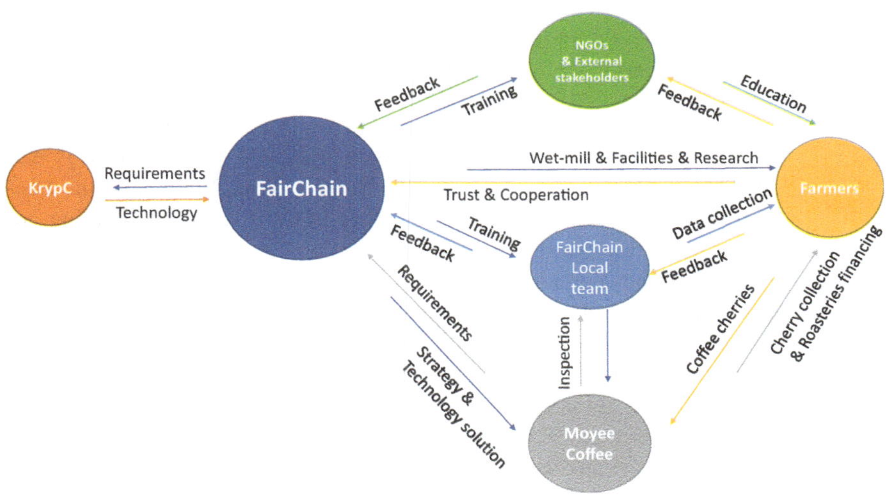

Fig. 6.9 Stakeholder relationship (Source: the author)

6.3.3.1 Farmers' Engagement

To engage farmers, FairChain started to build a close relationship with farmers by providing incentives and motivations. To achieve this, FairChain took certain actions:

1. Increase purchasing price to attract farmers

   The first thing FairChain did to motivate farmers was to increase the cherry collection price. "*So in the beginning you had to find farmers who want to supply cherries to you. Of course, pay a better price always help. But they're now just lining up, they want to be part of this FairChain farmer programme*", said by Mr. R. Therefore, farmers who sell cherries to FairChain can receive 20% more than the market price: 10% was paid directly to the farmers and another half was paid into the farmer representatives' saving accounts to cover the education fee (FairChain Foundation, 2019a). With the high purchasing price, more farmers are willing to sell their cherries. In addition, six farmer representatives were chosen to take care of farmers for FairChain, particularly during harvest (FairChain Foundation, 2017). The farmer representative is responsible for quality control during harvest season.
2. Gain trust and build a close relationship with farmers

   FairChain gains trust and builds a strong relationship with farmers. Increasing price only is not enough to solve the problem from the root. Therefore, it is important to let farmers know what technologies can do for them. However, according to Dr. Deepak, a researcher at the University of Leeds, who has investigated FairChain, "*I have identified some potential challenges from the FairChain case, which is that people who are using blockchain technology do not have knowledge about it. So, lack of understanding of the technology is one of the major challenges especially at the lower tier level*". Therefore, FairChain has provided multiple training sessions to farmers to target different aspects. For lack of understanding of technology, FairChain firstly educated NGOs and local team about blockchain, then farmers can receive training from NGOs and the local team. For example, 15 lead farmers are selected to take part in the training and lead by example. They are responsible for educating other farmers agricultural knowledge (FairChain Foundation, 2019b).

   FairChain also paid close attention on educating younger generations as the interview shows that they are more eager to learn new things (Dekker et al., 2019). For production development training, an "Outgrower program" was launched to provide agricultural knowledge about how to double the yield. 2.5%–15% of the farmers (362) who have participated the training in 2018 had adopted new skills (FairChain Foundation, 2019a). Financial empowerment programme was conducted one to one to educate farmers about technology and banking system (FairChain Foundation, 2020). The digitalisation will allow farmers to participate more in the future programmes such as microloan programme. The farmer education can significantly enhance the relationship and trust between FairChain and farmers, as well as gaining knowledge.
3. Re-distribute Value in the Coffee Supply Chain.

   Rather than simply paying a bit more to local producers, FairChain is thinking even further. They have moved the value-added processes (roasting & packing) to the country of origins. In order to do so, FairChain financed a wet mill to the local farmer community in 2018. This allowed coffee beans to be washed after cherry collection, and farmers can get extra profit from it. Moreover, a roastery in Kenya was established in 2020 to add more value to the products (Moyee Coffee, 2021). The new facilities not only retain more value in Ethiopia but have also provided more job opportunities to local people. More importantly, over time, the ownership of these facilities will be shared by all stakeholders, including farmers and consumers. As Mr. R said, "*that's why we said shared value chain, it is not about us owning everything and just buying from the farmer. Now, the ambition is that all stakeholders in the value chain, including the consumers, will be owning our value chain of coffee. That's what we are now working on*".

6.3.3.2 Consumer Involvement

The consumer is also an important factor to promote value distribution in the project. Apart from showing every transaction to consumers, FairChain has more ambition. It has converted its product marketing budget to digital loyalty tokens (Stokes & Reilly, 2019). "*So instead of spending the marketing budget for the brand in Facebook or Snapchat or you know, an advertisement running online or offline, but put that marketing money into the hands of the consumers, so whenever they purchase a cup of coffee, they receive a token*", said Mr. A. The tokens can be activated by scanning the QR code on the product package, to redeem 25–50 cents discounts for next purchase or reinvest to the farmers for tree planting (tip the farmer). How to use the token is fully dependent on consumers. "*Our hypothesis is that if you can bring a fair traceable product to market and you can trigger the consumer to do something good, then consumers are most likely to do it instead of spending 50 cents on discount*", explained Mr. A.

6.3.3.3 Impact Consortia

Other stakeholders who are willing jointly create impact are also making efforts in this case. The office of the FairChain Foundation is based in the Netherlands and shares different language and culture with farmers in Ethiopia. Meanwhile, farmers are suffering high poverty rate (90.70% in 2015) and low literacy rate (51.77% in 2017). To conduct farmer training, it is necessary to have local communities and local teams to connect farmers and FairChain like a bridge. Thus, governments (Dutch embassy in Ethiopia), NGOs (South Africa), and the local university (Jimma University) are needed to communicate with farmers. "*Moyee sells the coffee, we provide the digitisation as FairChain tech. But I also need NGOs like South Africa or FairChain foundation local team in Ethiopia to execute the programme from the ground, because you cannot run an impact programme from Amsterdam where the beneficiaries are in Ethiopia, that doesn't work*", said Mr. R. Therefore, an "impact consortia" between multiple organisations was created to achieve the same goal—bring farmers to a living income. The external stakeholders are not part of the value chain as they are only responsible for assisting and promoting the project. Therefore, they are not on the blockchain platform.

6.3.4 Future Plans

FairChain is contacting PwC and planning to do continuous auditing. Instead of the annual impact report, auditing companies can do continuous auditing with blockchain. According to Mr. R, "*that's not in place yet, but that's one of the positive side effects of running this blockchain and that you're able to do a continuous audit thing instead of once-a-year impact report auditing*". The continuous auditing not only improves efficiency and saves time, but also increases the credibility and legitimacy since the data are updated near real time and are immutable.

FairChain is planning to go further in sustainability. For example, the Microloan Project is coming soon. It is a project where consumers can buy more tokens and lend to farmers to expand their business. Farmers can pay back the consumers after each harvest season. An economic cycle of growth is expected to be achieved by the Microloan Project. Another example is the Women's Empowerment Committee, which is going to be formed to motivate women and integrate female power. From the interview results by FairChain (Dekker et al., 2019), females are less educated in Ethiopia. Therefore, FairChain hopes to encourage women to join the project and improve their living conditions. Moreover, FairChain is exploring the possibilities of recycling coffee product waste for another use to increase farmers' incomes. FairChain is also planning to operate the wet mill in solar power to reduce the environmental impact from coffee production.

FairChain aims to expand its products' scale; more products that are empowered by blockchain will be on the market soon such as chocolate and honey projects. These products are similar to the Moyee coffee project, in that the products will be traceable on blockchain and aim to promote an equally shared value chain. Eventually, FairChain hopes to become a lead example in the industry to promote sustainability.

FairChain is working to build the system on a public chain. Compared to permissioned blockchain, the public chain will be accessible to everyone to check information. According to Mr. R, "*if our claim is that we contribute to living income and we improve the living income, you should be able to check that claim without asking for access to my data or something like that. And that's what we're currently working on with the team on how to make this public infrastructure available for all our consumers*".

6.4 Blockchain Benefits

This section summarises the benefits that blockchain has brought to FairChain, and how blockchain addressed the issues mentioned in Sect. 6.2. The issues cannot be addressed without the transparency in the supply chain supported by blockchain.

6.4.1 Story Proving

FairChain is aiming to restore the coffee value chain balance back to 50/50, therefore, Blockchain in this case is used as a story-proving tool to show the public how FairChain helped to achieve the shared value chain goal. "*For us, it is a pure provenance of the product value distribution, we are fully focusing on a quantifiable, verifiable proof of impact. I am running this on living incomes, and we digitise the value chain. So, the digitalising is a mean, not an objective*", said by Mr. R.

Decentralisation and Immutability are the unique features of blockchain which can bring transparency and traceability to the coffee supply chain, to trace coffee from seed to cup. From the farmer in Ethiopia to the warehouse in the Netherlands, every transaction is updated on the blockchain platform in real time. While every stakeholder has the same power to check the information, the immutable record also ensures the data authenticity. Therefore, blockchain can be considered as evidence to support what FairChain said it achieved, because the public can check the outcome themselves on blockchain and trust the technology.

This is particularly usefully when there is limited transparency in the supply chain and consumers have minimal trust on what companies say. "*So what we want to see running backed up by blockchain is that the value distribution is clear for our consumers, because for us, it's important that we can prove 50% of the value of the coffee as well. We sell a product, a bag of coffee for one kilogram, for 20 euros, something like that. And we want to demonstrate to our consumers that 50% of that remains in Ethiopia. And that's why I use blockchain to have this immutable proof*", said Mr. R. The two issues that are mentioned above—value inequality and sustainability issues—can be proved under the radical transparency that blockchain brought.

6.4.1.1 Shared Value Chain

The first thing FairChain hopes to prove is their efforts on restoring coffee value. In order to do so, FairChain proposed a "trade over aid" concept, which means that they hope to help coffee producers by improving product quantity and quality. Therefore, in addition to paying 20% more than market price to farmers, FairChain also arranged a few training sessions (agricultural, financial, and technological training) for farmers to improve their knowledge on coffee production and their acceptance of new technologies. Moreover, FairChain also invested facilities to provide job opportunities to local areas and help them to gain added value from activities such as cherry processing and beans roasting. Therefore, rather than simply increasing prices, FairChain is recreating the local business model—from earning a basic salary from tree planting to earning more added value by more business activities.

Everything that FairChain did can be eventually checkable on the blockchain platform. By scanning the QR code, consumers can see who grows the tree, where the coffee bean has been processed, the grading score, roasting condition, when it has been exported and more importantly, the costs in each step. Different from other use cases, cash flow in this project is an important part to show to consumers, because it is the evidence to prove that at least 50% of the value remains in Ethiopia. Information of key stakeholders is also provided. This will provide an insight of the value chain; for example, consumers will get to know exactly who they are helping. The transparency will not only enhance trust between consumers and FairChain but can also be used for targeted poverty reduction. Coffee producers can also tell their stories to consumers and build a longer and stronger relationship with them. "*That's why we call it radical transparency. The consumers will know in the value chain about who owns what. We've got the supply chain solution, which basically provides provenance of the product. And it also makes you follow the money. And you can look back from your product that you bought to who did earn what*", Mr. R explained.

6.4.1.2 Positive Externalities

Another thing FairChain wants to prove is the positive externalities—sustainability. Sustainability has been mentioned more often in recent years, and consumers also tend to consume more sustainable products (Ryan, 2019). However, when every company is saying their products are sustainable, it is hard to trust only based on what the companies say. Although there are many sustainable certificates such as Fairtrade and Rainforest alliance, in many cases, sustainability is only used as way of marketing rather than actual behaviour. For example, according to Truthinadvertising.org (2020), many companies are accused of Greenwashing—the products are not environmentally friendly as the company asserted. The Greenwash company list includes world-leading food company—Nestle, that was accused in 2019 of deforestation and use of child labour force for its "sustainable sources coca beans". Therefore, it is important for FairChain to gain trust from consumers.

Apart from trusting on the sustainable certificates, consumers can now trace what FairChain has done on blockchain. More importantly, the data are authentic and cannot be tampered with on blockchain. For example, the facilities that FairChain invested in can make many value-added activities remain in the country of origin. Therefore, consumers can see that their coffee has been processed, roasted, and packed in Ethiopia, and the coffee has been exported to Europe in a packed condition. More specifically, consumers can not only check the costs in each step and know where their money has been spent, but also can see the product flow and know value-added activities have been completed in the country of origin. The whole supply chain is just as FairChain claimed. Blockchain can prove that FairChain is telling the truth about the shared value chain and sustainability.

6.4.2 *Marketing*

Another benefit of blockchain is to involve consumers in the new system. In this project, consumers can do more than information checking; they can also become part of the shared value chain. Apart from purchasing Moyee coffee products, consumers can also activate tokens and tip the coffee farmers. The tips can then be used to plant new trees and add more value to the value chain. Between consumers and farmers, a closed loop of value is formed.

In addition, the consumer engagement can also enhance their brand loyalty. Consumers can see how their support can bring improvements in Ethiopia. Like Mr. R said, "*we have a token for a tree, the tree contributes to living income over years, because it's close to the harvest as I guess more cherry. And it also captures a bit of carbon dioxide. So it's also a positive externality of capturing carbon, and then over time that's what we can demonstrate to our consumers, if you provide a token to a farmer, over time, you can see what that does with the living income*". FairChain has redistributed its marketing budget and create a direct link between consumers and farmers. The sense of participation will encourage consumers to engage more. According to Mr. R, "*we work on what we call a brand loyalty system. The idea behind this blockchain solution is that we try to connect with consumers in a different way. And we're running experiments to see if providing insights into the value chain and value distribution can do anything good for customer loyalty*".

6.5 Implementation Barriers

This section presents the actual barriers that FairChain encountered during blockchain implementation, and how they overcame them (Table 6.2). Barriers are divided into four perspectives based on Saberi et al. (2018): Intra-organisational barriers, inter-organisational barriers, system-related barriers, and external barriers.

Blockchain was chosen in this project to inform the public and to prove what FairChain has done on promoting the shared value chain. Both personal money and partial government funding were invested to promote the project. To cover the costs, FairChain also provided technology solutions to other companies to earn extra revenue. The strong determination of FairChain to bring farmers to an income level is also a driving force to promote the project.

In this project, farmers are suffering high poverty rate and low literacy rate. They do not trust either foreigners or local management systems. Therefore, FairChain had to gain trust and built a close relationship with farmers. FairChain began with increasing the cherry purchasing price and looked for help from local communicates and NGOs. The high purchasing price successfully attracted more farmers to join the project, and local communities and NGOs helped to connect farmers and FairChain. To enhance the relationship with farmers, FairChain arranged a few training sessions via NGOs and local communities to educate farmers in the new knowledge of

Table 6.2 Barriers from four perspectives

Four perspectives	Barriers' categories	Barriers in practice	Approaches and solutions	Supporting quotations
Intra-organisational barriers: the barriers in this category are from the internal of the organisation	Company demand and ambition	Technology choosing and keep strong determination	FairChain decided to use blockchain to prove their efforts and achievements in helping farmers. The shared value chain is both a company demand and a driving force for FairChain to use blockchain. Moreover, coffee supply chain is complex and broken. As a non-perishable product with added value, the coffee value chain has been very unequally distributed. Therefore, there is a large value to use blockchain to trace costs and profits and provide transparency. Moreover, coffee is one of the most accessible drinks for consumers.	"Guido van Staveren found out some research that the coffee value chain is a very unequal disputed value chain. So he said, okay, I'm going to build this FairChain foundation with the objective to build shared value chain with the positive externalities. And so instead of all the negative externalities like carbon footprint, we want to create positive externalities. So, that's why he started. And the first coffee was brought to Europe, I think, 2015 or something like that. And then he just brought a coffee and paid a bit more. And we found out that we needed some kind of solution to trace it, or we can do story telling".—Mr. R, the chief technology officer of FairChain Foundation. "So far blockchain was used for non-perishable products are

(continued)

Table 6.2 (continued)

Four perspectives	Barriers' categories	Barriers in practice	Approaches and solutions	Supporting quotations
				mostly value added products. Like for example, rough food, there's not much value addition, not much profit. The product characteristics, the supply chain situation is different".—Dr. Deepak, a researcher at the University of Leeds "From global supply chains is different; consumers don't actually have the access to what happens elsewhere".—Dr. Deepak, a researcher at the University of Leeds "By starting with food is because it's the most accessible for consumers. The only thing that you have to do is to create a better life for a farmer by buying a coffee, which you are most likely doing several times a week in your natural behaviour. You don't buy a luxury watch a couple of times a week. You do it by coffee a

(continued)

Table 6.2 (continued)

Four perspectives	Barriers' categories	Barriers in practice	Approaches and solutions	Supporting quotations
				couple of times a week. So our model is based on changing existing consumer behaviour by, or not changing it, leaving consumer behaviour as it is, and then providing them with another step in which they can make more impact".—Mr. A, the FairChain Architect.
	Financial constraints	Massive initial investment	Half of the investments are personal money, and partial are from government funding. Moreover, FairChain also provides technology solutions to other companies based on a cost-plus model to earn extra cash.	"So the cost of for development of the platform is partially private money and partially governmental funding from the Dutch government. For running the systems, that's why we have to have the FairChain entity. We sell software solutions to the market on a cost-plus model. So I have a list of applications we have and the list of tools we can provide to the market and our consumer".—Mr. R, the chief technology officer of FairChain Foundation. "Seeing return on investment, nobody really knows. How

(continued)

Table 6.2 (continued)

Four perspectives	Barriers' categories	Barriers in practice	Approaches and solutions	Supporting quotations
				much is that? There's always a fear. Like every time there's a new technology comes in, there's always a fear to invest in this new technology".—Dr. Deepak, a researcher at the University of Leeds
Inter-organisational barriers: the barriers are between the supply chain stakeholders	Stakeholder engagement	Lack of incentives for farmers to participate	For farmers, FairChain paid 20% more than market price in order to attract their interest to join the project, as well as build a strong relationship with them.	"So in the beginning you had to find farmers who want to supply cherries to you. Of course, paying a better price always helps. But they're now just lining up, they want to be part of this FairChain farmer programme".—Mr. R, the chief technology officer of FairChain Foundation. "For farmers there is a big incentive. They get higher prices because the business model of FairChain is based on higher prices for farmers. There's long-term benefit as consumers might plant the tree and farmers one day wake up and receive trees from consumers".—

(continued)

Table 6.2 (continued)

Four perspectives	Barriers' categories	Barriers in practice	Approaches and solutions	Supporting quotations
				Mr. A, the FairChain Architect.
			FairChain bought a wet mill to the community (2018) to encourage value-added activities in the country of origin.	"And so we bought this this wet mill for the community. And so basically we financed it. We bought it and over time the farmers and the stakeholders in the FairChain, they become owners of this wet mill as well. We leave value added activities in Ethiopia, like roasting and packing, etc. So that creates jobs and that creates income at the end of the line".—Mr. R, the chief technology officer of FairChain Foundation.
		Lack of knowledge and distrust on local management systems	Multiple training, such as "Outgrower program" to provide farmers agronomic training, and financial training to help them open bank accounts and use smart wallets.	"I have identified some potential challenges from the FairChain case, which is that people who are using blockchain technology do not have knowledge about it. So lack of understanding of the technology is one of the major challenges particularly at the lower tier level".—Dr. Deepak, a researcher at the

(continued)

Table 6.2 (continued)

Four perspectives	Barriers' categories	Barriers in practice	Approaches and solutions	Supporting quotations
				University of Leeds "They don't need to understand blockchain technology, but they need to understand the value of the data that they feed into the system. Otherwise it's, it's more like a garbage in, garbage out and yeah. So, they need to have some kind of initial training to use the technology because everyone has to use this technology"—Dr. Deepak, a researcher at the University of Leeds "We worked with NGOs that we train in a platform use in doing, in the use of technology and the education of farmers. And that's going quite okay".—Mr. A, the FairChain Architect.
		Concern over personal privacy	Farmers are anonymised on the chain, and sensitive data are stored off-chain	"The farmer data is not in the blockchain itself. It's all anonymised. So a farmer has just a number in the blockchain it's called a farmer number something, and we've

(continued)

Table 6.2 (continued)

Four perspectives	Barriers' categories	Barriers in practice	Approaches and solutions	Supporting quotations
				got an off-chain database where, or a farmer profiles are in. And so part of the data is in the blockchain itself for immutability, the part of it is like the personal and privacy and GDPR-related data is in an off-chain database and not accessible for anyone".—Mr. R, the chief technology officer of FairChain Foundation. "So supply chain actor privacy is guaranteed by only storing transactional data on the blockchain. We don't source sensitive or personal data on the blockchain".—Mr. A, the FairChain Architect.
		Cultural and geographical difference	FairChain has invited NGOs, local communities, and local universities to communicate with and educate the farmers.	"We don't deploy those services ourselves. We work with NGOs that deliver that training. We train the NGOs in that sense. Because it is not a one-time practice. It's an ongoing process year around. So we cannot control that from an

(continued)

Table 6.2 (continued)

Four perspectives	Barriers' categories	Barriers in practice	Approaches and solutions	Supporting quotations
				office in Amsterdam. So we worked with NGOs that we train in how to use the technology and the education of farmers. And that's going quite okay".—Mr. A, the FairChain Architect.
		Lack of incentives to attract other stakeholders on board	For other stakeholders, only value chain-related stakeholders are required to upload costs and payments, other stakeholders do not need to disclose sensitive information. Moreover, FairChain has looked for companies that share the same sustainability goal with them.	"Top management challenges about convincing stakeholders to implement are always there".—Dr. Deepak, a researcher at the University of Leeds "It's also a part of a selection of partners. So, we say, we want to be transparent about the cost and the things we do. And we select companies that also want to comply to that, so that it's not that difficult. Our objective is to ensure that everybody makes a proper income out of our value chain and people understand that. So, we don't need to find people, they are knocking on our door".—Mr. R, the chief technology

(continued)

Table 6.2 (continued)

Four perspectives	Barriers' categories	Barriers in practice	Approaches and solutions	Supporting quotations
				officer of FairChain Foundation. "When companies do have secrets on their profits, the question that actually should arise is, why should there be a secret? Why should you hide that you are making a huge profit from just marketing and you are buying as low as you can? If you are a company that just does that, then blockchain is not the technology for you because blockchain will expose your current business. So, the company that is ethical will have much more interest in blockchain because they can prove what they already did".—Mr. A, the FairChain Architect.
		Lack of consumer awareness and engagement	Encourage consumers to activate token and reinvest to farmers for tree planting or for discount.	"So instead of spending the marketing budget for brand into Facebook or Snapchat or you know, an advertisement running online or offline, put that marketing money into

(continued)

Table 6.2 (continued)

Four perspectives	Barriers' categories	Barriers in practice	Approaches and solutions	Supporting quotations
				the hands of the consumers, so whenever they purchase a cup of coffee, they receive a token or, whenever they buy a chocolate bar, there's, it's open inside of the chocolate bar. Unlike a coin, you know. And once they scan that, they activate the marketing budget. So 25 cents or 50 cents off the brand and they can either spend it on this discount for themselves or they can share it towards such an impact goal such as the tree planting and NGO".—Mr. A, the FairChain Architect.
System-related barriers: the barriers related to the technology	Technology immaturity	Immature technology	Scalability issues	"I don't see scalability is a problem. Scalability is really, or replicability is always a short-term problem. it's literally the same thing as 10 years, 15 years ago you couldn't attach 500 kilobytes to an email and if you wanted to upload a photo, then it was three megabytes and you couldn't do that. So, I think

(continued)

Table 6.2 (continued)

Four perspectives	Barriers' categories	Barriers in practice	Approaches and solutions	Supporting quotations
				innovation will always lead before scalability".—Mr. A, the FairChain Architect.
		Data mistakes	FairChain has adopted IoT devices to help in data collection. In addition, FairChain has invited its local team, Moyee coffee to check data before data uploading.	"The big solution to this is called IOT, internet of things. So the smart scales and smart sensors lead the way to prevent intervention or the human touch towards entering data".—Mr. A, the FairChain Architect. "We have the local team on the ground who does the job, they check their own job. Moyee does a coffee as our partner in this coffee collection, they are the contact party with the farmer and they check that what the local team does is right. And FairChain is the third and final step after there's sort of a triple check on is the data correct. If all the transactions are nicely made before we publish them on the blockchain, that's why we use this triple accounting.

(continued)

Table 6.2 (continued)

Four perspectives	Barriers' categories	Barriers in practice	Approaches and solutions	Supporting quotations
				We validate that we have proof for harvest collection, we validate with the payments slip, and with the bank that the money is actually being transferred to. You know, I have a triple account. There's also a triple accounting mechanism in blockchain. We rely on the multiple sets of eyes checking data before it's entered in the bucket".—Mr. R, the chief technology officer of FairChain Foundation.
		Limited function	Blockchain itself cannot do everything. It normally requires a combination of other technologies to achieve better efficiency, such as IoT device and artificial intelligence (AI).	"I think that blockchain is not the Holy Grail, but the combination of blockchain with those IOT machines and machine learning and AI systems are bringing the whole package towards the level that it needs to be. So, it's not a substitute for blockchain. I think blockchain is one of the key components because it's just a ledger that is distributed in a way that everybody

(continued)

Table 6.2 (continued)

Four perspectives	Barriers' categories	Barriers in practice	Approaches and solutions	Supporting quotations
				can just look at the same information. And I think that is absolutely key to have. I think the infrastructures of blockchain will have to improve and also in regards to privacy. I do not think that there is a substitute for blockchain, but I think that only having blockchain is not good enough. It doesn't cover everything you need to work in collaborating and bringing other technologies in that it is only very distinct by name. But on a tech level, it is very much connected".—Mr. A, the FairChain Architect.
	Environmental restrictions	Digitalisation	Combine online and offline data collection methods together to avoid poor Internet connection. Data can be collected offline but will be synchronised and uploaded automatically when the network is back to normal.	"You cannot rely on online systems. You have to have sort of offline/semi-online mechanisms in place, and that's what we have for everything. And we can enter data without internet connection. And as soon as the internet is up

(continued)

Table 6.2 (continued)

Four perspectives	Barriers' categories	Barriers in practice	Approaches and solutions	Supporting quotations
				again, and we synchronise the data and do the checks, checks, and checks, and we send it through to blockchain".—Mr. R, the chief technology officer of FairChain Foundation.
		Digital literacy	Bring and encourage the new generations into the project.	"And then there's also an important challenge we have to deal with—digital literacy. In the field it is definitely a challenge. And so where the older farmers have families, the members of the farmer sometimes have issues with reading and writing. Luckily the new sort of generation and the sons and daughters of the farmers are on their way. And that's also a big opportunity because they know about smart phones, they know about technology, they know what technology can do for them. So, they are very eager to join us because they see there's a huge benefit of running things on blockchain.

(continued)

Table 6.2 (continued)

Four perspectives	Barriers' categories	Barriers in practice	Approaches and solutions	Supporting quotations
				Those challenges become an opportunity when the next generation comes into the blockchain".—Mr. R, the chief technology officer of FairChain Foundation.
			FairChain has improved its interface design to make the application more convenient for farmers. For example, the listen function allows people who cannot read to listen to the content.	"And so there are some very fundamental issues we have to deal with. So, we have to have an assistant model because farmers cannot read and write".—Mr. R, the chief technology officer of FairChain Foundation.
External barriers: the barriers from external environment.	Government policy, encouragement, and external stakeholder involvement	Lack of support from government and external stakeholders	FairChain has applied for some funding from the Dutch government to build the FairChain Foundation. Moreover, many NGOs and universities are involved to help to educate farmers.	"We work with the universities and some of those kinds of NGOs. We work with the Dutch government, Ethiopia, the Dutch embassy in Ethiopia. So, there are more people on that".—Mr. R, the chief technology officer of FairChain Foundation. "Governments are always behind in innovation. That's not the problem. Innovation happens. And governments see it happening.

(continued)

Table 6.2 (continued)

Four perspectives	Barriers' categories	Barriers in practice	Approaches and solutions	Supporting quotations
				They see it failing, they see it flourishing and they provide the regulatory frameworks to scale it and make it responsible".—Mr. A, the FairChain Architect.

agriculture, technology, and finance. In addition, FairChain also financed a few facilities to the local area to help farmers to earn more added value and provided more jobs in Ethiopia. Consumers are also invited to join the value chain by sharing tokens with farmers. The tokens can be used to reinvest for tree planting and bring more value to local area. Considering personal privacy, farmers' sensitive information is stored off-chain; only the value-related information is on the chain. Apart from stakeholders in the value chain, other external stakeholders are neither required to be on chain nor provide sensitive information.

In the technology level, poor Internet connectivity, low smart phone possession rate, and data mistakes are the main challenges. FairChain came up with a combination of online and offline methods to overcome the poor network issue. Data can be captured by IoT devices offline when there is limited network but will be automatically uploaded on the chain when the network becomes strong. This not only increases data accuracy and prevents human intervention, but also solves the issues with limited network connectivity. The interview FairChain conducted in mid-2019 suggests that young generations tend to be more passionate about new technologies. They have deeper understanding about how technology will change their lives compared to the older generation. The new generation is an opportunity of FairChain to promote the project and bring more positive externalities to the local area. Moreover, FairChain also tried to improve the user experience by adding some new functions to make the APP easier for farmers, such as the listen function for people who cannot read. To avoid data mistakes, the triple accounting mechanism was built by FairChain local team, Moyee coffee, and FairChain. Before uploading data, Moyee coffee and the local team will first double check the data authenticity and FairChain will have a final check to allow data imputing.

External stakeholders in this case are the facilitators who jointly create a positive impact for Ethiopia. They have helped significantly in this project. When FairChain was built, partial funding was from the Dutch government. During blockchain implementing, local communities and NGOs were there to help FairChain with communication issues. Every step FairChain has promoted this project, external stakeholders were involved to contribute to it.

6.6 Case Summary

Overall, FairChain was founded with the goal to restore the coffee value chain. Since 2013, FairChain has worked with NGOs to promote the sustainable business plan, including finance facilities, provide training, and increase purchasing prices. Blockchain was tested since 2017; later, it was officially adopted to enable tokenisation and to improve transparency and credibility. Blockchain, as a story-proving tool, can show the public what they have achieved so far. By using blockchain, consumers can not only check the products and costs in every step along the supply chain but can also get involved in the value chain by activating and investing their tokens. This chapter has introduced the background of FairChain, the issues in the coffee supply chain, the implementation stages, the benefits that blockchain has brought, and the barriers FairChain overcame.

References

Ashoka. (2014). 5 Things you need to know about your daily cup of coffee. *Forbes*. Accessed 18.02.2021, from https://www.forbes.com/sites/ashoka/2014/07/16/5-things-you-need-to-know-about-your-daily-cup-of-coffee/?sh=1be1240a7e69

Bruce-Lockhart, C., & Terazono, E. (2019). *From bean to cup, what goes into the cost of your coffee?* Accessed 15.02.2021, from https://www.ft.com/content/44bd6a8e-83a5-11e9-9935-ad75bb96c849

Businesswire. (2020). *Global coffee market (2020 to 2026) – industry perspective, comprehensive analysis and forecast – ResearchAndMarkets.com*. Accessed 15.02.2021, from https://www.businesswire.com/news/home/20201006005799/en/Global-Coffee-Market-2020-to-2026%2D%2D-Industry-Perspective-Comprehensive-Analysis-and-Forecast%2D%2D-ResearchAndMarkets.com

CBI. (2019). *Exporting coffee to the Netherlands*. Accessed 08.02.2021, from https://www.cbi.eu/node/721/pdf

CBI. (2020). *What is the demand for coffee on the European market?* Accessed 08.02.2021, from https://www.cbi.eu/market-information/coffee/trade-statistics

Coffee Bean Corral. (2021). How coffee is made? *Coffee Bean Corral*. Accessed 08.02.2021, from https://www.coffeebeancorral.com/HowCoffeeIsMade.aspx

CropIn Technology. (2020). *SmartFarm: A tailor-made digital solution for plantation crops*. Accessed 08.03.2021, from https://cropin-tech.medium.com/smartfarm-a-tailor-made-digital-solution-for-plantation-crops-417273b3ecb5

Dekker, S. D., Kust, M. J. C. V. D., & Rijcken, L. (2019). *Empower coffee farmers through mobile phone usage*. Accessed 09.02.2021, from https://mk0FairChainorgou7il.kinstacdn.com/wp-content/uploads/2020/02/Empowerment-of-coffee-farmers-through-mobile-phone-usage-Limmu-Ethiopia-July-2019-1.pdf

FairChain Foundation. (2014). *FairChain Coffee Moyee Wins Radical Innovators 2014*. Accessed 09.02.2021, from https://FairChain.org/moyee-wins-radical-innovators-2014/

FairChain Foundation. (2017). *Meet Shamshedin/farmer representative*. Accessed 09.02.2021, from https://FairChain.org/meet-shamshedin-farmer-representative/

FairChain Foundation. (2018). *Stories from Ethiopia, Ilse's Blog, part 1*. Accessed 09.02.2021, from https://FairChain.org/stories-from-ethiopia-1-2/

FairChain Foundation. (2019a). *FairChain insights*. Accessed 09.02.2021, from https://FairChain.org/FairChain_insights_1/
FairChain Foundation. (2019b). *FairChain Farming Whitepaper*. Accessed 09.02.2021, from https://FairChain.org/wp-content/uploads/2019/02/FC-Farming-Whitepaper.pdf
FairChain Foundation. (2019c). *x Blockchain for living income*. Accessed 09.02.2021, from https://FairChain.org/blockchain-info/
FairChain Foundation. (2020). *Digitisation: The next giant leap for global coffee communities*. Accessed 10.02.2021, from https://FairChain.org/bankaccounts/
Fairtrade. (2019). *We love coffee, are we willing to pay the price?* Accessed 08.02.2021, from https://www.fairtrade.net/news/we-love-coffee-are-we-willing-to-pay-the-price
Fairtrade. (2021) *Coffee*. Accessed 08.02.2021, from https://www.fairtrade.net/product/coffee
Field, A. (2014). A supply chain overhaul to boost coffee farmers' income 400%. *Forbes*. Accessed 08.02.2021, from https://www.forbes.com/sites/annefield/2014/07/16/a-supply-chain-overhaul-to-boost-coffee-farmers-income-400/?sh=6e1ab07b13b2
Knoema. (2021). *Ethiopia - Poverty headcount ratio at $3.2 a day based on purchasing-power-parity in constant prices of 2011*. Accessed 08.02.2021, from https://knoema.com/atlas/Ethiopia/Poverty-rate-at-dollar32-a-day
Kucukcay, I. (2020). *Coffee: A favorite beverage with a complex supply chain*. Kinaxis. Accessed 18.02.2021, from https://www.kinaxis.com/en/blog/coffee-favorite-beverage-complex-supply-chain
Macrotrends. (2021). *Ethiopia literacy rate 1994–2021*. Accessed 10.02.2021, from https://www.macrotrends.net/countries/ETH/ethiopia/literacy-rate
Moyee Coffee. (2021). *Single Origin*. Accessed 10.02.2021, from https://moyeecoffee.ie/collections/our-coffee/products/single-origin?variant=7455731613751
Rushton, D. (2019). *Map of the month: Bringing smallholder coffee farmers out of poverty*. Accessed 10.02.2021, from https://carto.com/blog/enveritas-coffee-poverty-visualisalisation/
Ryan, T. J. (2019). *Consumers willing to pay up for sustainability*. Accessed 10.02.2021, from https://sgbonline.com/consumers-willing-to-pay-up-for-sustainability/#:~:text=According%20to%20CGS%202019,pay%20more%20for%20sustainable%20products.&text=Gen%2DZ%20was%20found%20to,compared%20to%20other%20age%20groups
Saberi, S., Kouhizadeh, M., Sarkis, J., & Shen, L. (2018). Blockchain technology and its relationships to sustainable supply chain management. *International Journal of Production Research, 57*(7), 2117–2135.
Samper, L. F., & Quinones-Ruiz, X. F. (2017). Towards a balanced sustainability vision for the coffee industry. *Sustainable Supply Chain Management, 6*(2), 1–28.
Stokes, K., & Reilly, S. (2019). *How Moyee coffee is introducing blockchain to radically improve the lives of coffee farmers across the world*. Available at: https://www.blockchain-expo.com/2019/04/blockchain/how-moyee-coffee-is-introducing-blockchain-to-radically-improvethe-lives-of-coffee-farmers-across-the-world/. Accessed on: 10.02.2021.
Truthinadvertising.org. (2020). *Earth day 2020: Companies accused of greenwashing*. Accessed 13.02.2021, from https://www.truthinadvertising.org/six-companies-accused-greenwashing/).
UNCTAD. (2020). *Fair trade enterprises spread benefits through value chains*. Accessed 13.02.2021, from https://unctad.org/news/fair-trade-enterprises-spread-benefits-through-value-chains

Chapter 7
Cross-Case Analysis

This chapter conducts a cross-case analysis between the three case studies. The chapter starts with similarities and differences in the blockchain adoption process (Sect. 7.1); followed by the key success factors that influence blockchain implementation (Sect. 7.2); next the common barriers and solutions are summarised in Sect. 7.3; and the performance outcomes of blockchain projects are presented in the end in Sect. 7.4. The theories of the innovation process model (Flint et al., 2005; Su et al., 2011) and the practice-based view (PBV) are applied to analyse the blockchain implementation in the case companies.

7.1 Innovation Process Model

This section compares the different practices within the four stages of the innovation process of the three case companies (Table 7.1). The four stages are (1) setting the stage activities, (2) customer clue-gathering activities, (3) negotiating, clarifying, and reflecting activities, and 4) inter-organisational learning (Flint et al., 2005; Su et al., 2011, Fig. 7.1).

7.1.1 Setting the Stage Activities

7.1.1.1 Identify the Issues and Clarify Company Demand

It is important to identify current issues within the supply chain and clarify demands in the initial stages. Although all the three case companies are in the food industry, the issues they were facing and the company demands were different due to the different characteristics of the product and the supply chain.

C. Zhang et al., *Blockchain Applications in Food Supply Chain Management*, Contributions to Management Science,
https://doi.org/10.1007/978-3-031-27054-3_7

Table 7.1 Innovation process stages in details

Innovation process stages	Sub-process	Practices	BeefLedger	W Company	FairChain
Setting the stage activities	Identify the issues and clarify company demand	Key issues	1. Fragmented information flow and misinformation within the cross-border supply chain 2. Inefficient cross-border document and payment process 3. Consumer concerns about food authenticity and safety	1. Consumer concerns about food safety 2. Inefficient supplier management and waste of labour investment in auditing	1. Value inequality in coffee supply chain and poverty for coffee farmers 2. Unsustainable sourcing in coffee industry
		Demand and goals	1. Streamline the information flow within the cross-border supply chain 2. Improve the efficiency of cross-border document and payment process 3. Increase consumer buying confidence	1. Enhance consumer buying confidence 2. Improve the efficiency of supplier management	1. Storytelling and story-proving of what FairChain has done to achieve equally shared value chain and sustainability 2. Value redistribution and improve the positive externalities
	Identify partners	Business partners	Food Agility—Research group Queensland University of Technology (QUT)—Technology provider	PwC—Strategy and consultancy VeChain—Technology provider	Moyee Coffee—Coffee seller KripC—Technology provider NGOs and local communities—Training and communication to farmers
		Supply chain partners	Breeder, producer, abattoir and processor, exporter, distributor, e-commerce, local agent, consumers	Upper stream suppliers, Tier 1 suppliers, consumers.	Farmers, wet mill, grading, hulling, grading, roasting and packing, warehouse, retailer, consumers
Customer clue-gathering activities	Data collection and proposed business model	Information collection	1. Reflect on existing Industry experience 2. Conduct internal research	1. Proposed multiple business plans 2. Went to a few industrial conferences	1. Semi-structured interview on local household living level and digital literacy

			about blockchain and food supply chain		
	Run pilots	Pilot experience	Piloting with Food Agility and Queensland University of Technology	Studying from the past piloting experience at the global level	Piloting with blockchain technology provider
Negotiating, clarifying, and reflecting activities	Design and implement new system	System design and building	1. Adopting RFID technology 2. Develop multi-signature function 3. Tokenise the products 4. Constructing infrastructures in some areas	1. QR code for suppliers instead of costly RFID 2. QR code per item for selected product 3. Adopting QR code and scan to webpage instead of a dedicated app for consumers	1. SmartFarm app for data collection 2. Enabling tokenisation function for consumers to expand the positive externalities
		Infrastructure selection	Ethereum—private and alliance blockchain	VeChainThor—public blockchain	Hyperledger fabric—private blockchain
	Supplier engagement	Supplier engagement	1. BeefLedger employed professionals to check supplier acceptance to blockchain 2. BeefLedger provided education to suppliers and persuade them from the market perspective	1. Supplier conference to discuss blockchain 2. Promised better shelf place to traceable products 3. Suppliers have little financial investment 4. The system is designed to be user friendly without impact on production efficiency	1. Building a close relationship with farmers 2. 20% higher purchasing price than the market to farmers 3. Training and education to farmers
Inter-organisational learning	Stakeholder collaboration	External partners	Governments	Governments	Governments NGOs Banks
		Other stakeholder engagement	1. Holding a few road-showcases and conducting research about consumer value perceptions 2. Collaborating with	1. Inviting more suppliers to participate 2. Having collected legal	1. Receiving support from governments, NGOs, and local communities. 2. Contacting auditing

(continued)

Table 7.1 (continued)

Innovation process stages	Sub-process	Practices	BeefLedger	W Company	FairChain
			E-commerce and local agents for product selling 3. Having a dedicated legal department to solve policy issues	opinions to solve policy issues	companies and reduce audit frequency 3. Consumers are encouraged to contribute to the equally shared value chain by lending farmers more tokens to expand business in the Microloan Project.
	System adjusting and business adapting	System adjusting	1. Improving interface design to be more user friendly 2. Knowledge classification to provide consumers different products' information based on demand 3. Plan to apply smart contract for smoothing documentation and payment process 4. Providing blockchain solutions to other brands. 5. Exploring the potential of supply chain finance	1. System optimising for suppliers 2. System interface design improving to make it more consumer-friendly 3.Enabling smart contract for automatic payment 4. Involving more products and suppliers 5. Changing targeting groups from supercentres to Sam's club	1. Listening function for farmers who cannot read 2. Younger generations have higher digital literacy and are the great driving force to blockchain implementation 3. Moving from private chain to a public blockchain 4. Expanding product scales to cocoa products such as chocolates

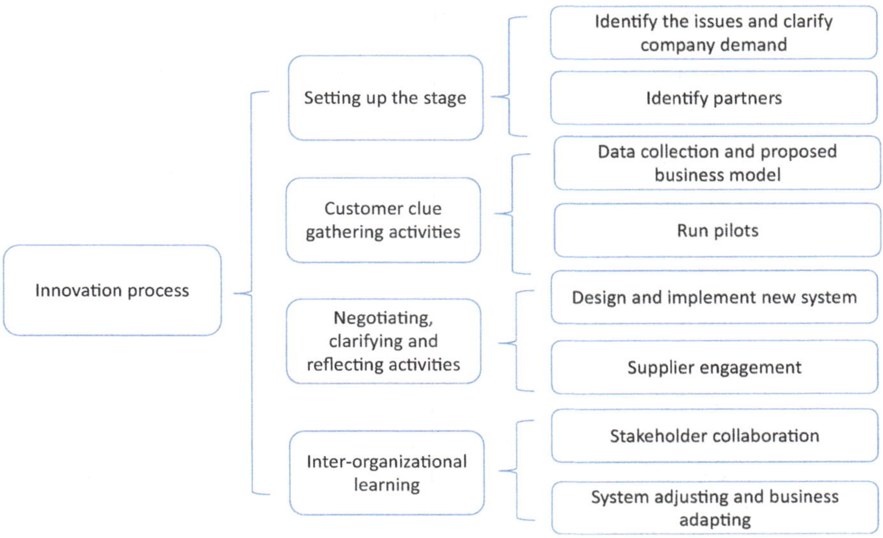

Fig. 7.1 Innovation process stages (*Source: the author*)

BeefLedger was facing issues of fragmented information flow, inefficient documents and payment process due to the cross-border beef supply chain. The issues are mainly caused by geographical, cultural, political, and time difference between stakeholders. When it comes to cross-border beef supply chain, the efficiency of information flow and document procedures, and the traceability of the entire chain are becoming more important. The general idea is the shorter the procedures of documentation, the quicker the product turnover, and the lower the operation costs.

Moreover, the product that BeefLedger is selling also suggests its unique product character—high quality and high price. The quality and reputation of Australian beef product also means higher average selling price than normal beef product. This not only attracts consumers to shop for quality Australian beef products, but also lures many counterfeiters in the market. Following many food scandals in China, Chinese consumers care more about food safety and authenticity in recent years. Therefore, another issue for BeefLedger is how to prove their product authenticity and quality and enhance consumer buying confidence.

One of the issues that W company was facing is similar to BeefLedger—consumer concerns. W company is an international well-known brand, and consumers tend to have more trust and faith in the large food retailers according to Kendall et al. (2019). Under the different situation with BeefLedger, consumer concern is still an issue for W company. One of the reasons is the consequences of the food scandals in the Chinese market, mentioned above. Chinese consumers are more cautious about food products. Another reason is that W company has more responsibility to provide better food and service to their consumers and secure their reputation in the meantime.

Apart from consumer concerns, W company has another unique issue—the nature of its supplier management. As a large retailer, W company has contracts with thousands of suppliers. To manage the large number of suppliers, it usually requires a huge amount of labour investment for factory inspection and auditing on both product quality and supplier compliance on corporate social responsibilities. The old-fashioned supplier management system is not only inefficient with waste of labour resources, but also allows possibility of data alteration and potential disputes. Therefore, W company has a need for an innovation that can efficiently manage supplier data and reduce risks.

In contrast to BeefLedger and W company, as a social enterprise, FairChain is concerned with how to prove what they did in achieving an equal coffee value chain and sustainability. In the FairChain case, food safety is the least important concern for FairChain concerns. This is because coffee is neither perishable nor expensive. Therefore, FairChain is less concerned about information flow and the procedure time within the supply chain. FairChain was initially founded with an aim to re-distribute the coffee value chain and promote a sustainable coffee supply chain. However, many food companies are using sustainability as a marketing strategy rather than putting it into practice. The main demand for FairChain adopting the new technology is to show the public what they have achieved. What FairChain needs is a power tool that can differentiate them from other companies by story-proving. Moreover, FairChain also hoped to encourage consumer participation to promote positive externalities.

After identifying key issues, demand and goal need to be established. To solve issues with consumer concerns, BeefLedger and W company have similar requirements on the innovation—real-time information flow and immutability. The real-time information flow can reduce the information asymmetry and increase transparency. This function is particularly useful for BeefLedger's cross-border supply chain. Immutable record can be used for food recall and define liabilities. It can also encourage companies and suppliers to behave legitimately. Both real-time information flow and immutability features can also be used for story-proving in the case of FairChain. The innovation that can provide immutable records and transparency to stakeholders will enhance FairChain's credibility. Proving what has been done, therefore, is the key demand for FairChain.

To solve issue related to documentation and payment processes, BeefLedger hopes to adopt an innovation to digitise and speed up the process. Decentralisation, smart contract, and immutability are the key functions required. Decentralisation allows users to have equal power and share same information that is secured by immutability. The smart contract can be automatically executed once the conditions are met. Therefore, the combination of these features can largely reduce the time wasted in checking documents and certificates. Moreover, decentralisation and immutability can also meet the demands of supplier management in the case of W company.

Tokenisation is also needed for BeefLedger and FairChain to solve different demands. As an asset of value, the value that token carries is different from BeefLedger and FairChain. BeefLedger uses the tokens to bound with beef products to prove the uniqueness and authenticity of the products. This can combine with immutability to secure food authenticity. For FairChain, the tokens are used to engage consumers and maximise positive externalities. Rather than ending up with buying a coffee, consumers can activate tokens to create a value loop that return profits to farmers.

7.1.1.2 Identify Partners

As blockchain is a multi-stakeholder platform, it is essential to identify the relevant partners on-board. Thus, in the first stage, the three case companies identified and gathered key stakeholders. Before BeefLedger was officially founded, the Chairman of BeefLedger—Warwick and the co-founders had long experience in the beef industry. The experience brought Warwick deep insights of the industry issues, and key connections with industry experts including beef producers and processors. Since blockchain has gained attention, the idea of using blockchain to overcome the supply chain issues has also emerged. Meanwhile, the research company—Food Agility, and QUT have also shown interest in the BeefLedger project.

W company did the same before launching blockchain. PwC and VeChain were invited to discuss blockchain in food retailing. PwC has in-depth industry experience in retailing and can provide strategies for W company. VeChain which has blockchain implementation track records is responsible for providing technologies and putting the ideas into practice.

FairChain has partnered with Moyee coffee since it was founded. FairChain also coordinated with NGOs and local communities to communicate with farmers. With the help from local groups, FairChain can overcome cultural differences and language barriers.

7.1.2 Customer Clue-Gathering Activities

7.1.2.1 Data Collection and Proposed Business Model

After clarifying the company needs, the next step is to conduct research and collect useful data. The data can be from either consumers, technology, suppliers or industry perspectives in order to bring a whole view of the innovation. All three case companies have collected data before innovation adoption, but in different aspects. At this stage, a feasible business model is proposed.

For BeefLedger, information about technology has been thoroughly collected. A year-long research study about blockchain was conducted including blockchain benefits and drawbacks. This is because the founders of BeefLedger had long working experience in the beef industry. Therefore, the focus for BeefLedger was mainly on exploring technology potentials.

In the case of W company, the American branch already had two pilots from 2016. W company has gained a certain level of knowledge and experience of blockchain technology. However, the Chinese market is different from other food markets. Chinese consumers have stronger concerns over food safety due to the food scandals. Therefore, the China branch of W company has put more efforts into investigating the Chinese market trend. Food expert—PwC was invited to provide industry knowledge to W company. VeChain also joined the project to give technology consultancy.

The household level of the coffee farmers is the main information FairChain has researched. FairChain was founded with the goal to bring farmers to a living income and to promote sustainability. It is important to know the current living conditions of households, the reasons for low living income, and what farmers want. From the semi-structured interviews, FairChain obtained a clearer image of the farmers' living conditions. Based on the results of the research, FairChain created a "living income road map" which contains different ways to practically help farmers.

7.1.2.2 Run Pilots

Once the case companies have sufficient knowledge of the technology and the industry, the next step is to run pilots based on the business model to gain practical experience. All three companies went through the piloting stage before they officially launched their blockchain platforms.

From 2018 to 2019, BeefLedger teamed up with Food Agility, QUT, and a truck insurer—NTI to explore the potentials of blockchain in the cross-border beef industry. They conducted a pilot study to track the beef products from Australia to China. The goal of the pilot is to minimise the risks of counterfeit products. In the meantime, the supply chain process can also be monitored. From this pilot, BeefLedger gained a deeper insight of how blockchain can benefit food safety in a cross-border supply chain.

W company also conducted multiple test runs from 2016. The American branch traced mango and claimed that the tracing time fell from 7 days to 2.2 s. The China branch tracked pork from farm to fork and found out that blockchain brought significant efficiency in digitalising the certificates. Moreover, the pilot also ensured the details of the farm, such as temperature, expiration date, shipping information, and processing time. W company's partner—PwC also had experience in piloting blockchain traceability. Another business partner—VeChain had already worked with a few food brands to trace products. Before the blockchain project was kicked off, W company and its partners had proposed and tested multiple business plans to find the best one that fits the company demand.

FairChain also partnered with Bext360 to bring transparency to the coffee value chain. The information of origins, coffee quality, and real-time payment to the farmers is traceable and immutable for all stakeholders. The transparency can prove that FairChain has paid the living income to the farmers.

7.1.3 Negotiating, Clarifying, and Reflecting Activities

7.1.3.1 Design and Implement New System

The knowledge and understanding of blockchain from the second stage set the solid ground for this stage. The practical experience gained from piloting can be reflected in the system design. All the companies have designed the blockchain system and enabled different functions based on their demand, knowledge, and experience. They also selected different data collection methods to suit different needs.

Blockchain System Design

BeefLedger built the blockchain platform on top of Ethereum's infrastructure, which has a solid foundation and allows BeefLedger to build a second layer on top of it. Once the platform was built, BeefLedger enabled a multi-signature function to ensure data authenticity. Data can only be uploaded and saved on blockchain if all nodes agree on it. This function, at the same time, encourages users to upload responsible data. The function of tokenisation was also enabled to ensure the authenticity and uniqueness of the products. The combination of multi-signature and tokenisation can satisfy the demand for ensuring consumer buying confidence.

W company's blockchain platform is built on VeChainThor, which is a public blockchain infrastructure created by VeChain. Both W company and suppliers can trust the data on the chain, as public blockchain enhances the data transparency and data security.

Compared to the pilot, FairChain also enabled tokenisation in this stage, but to meet different demand. Token is the bridge that connects consumers and farmers. A closed value loop can be created once consumers tip the tokens to farmers. This can encourage consumers to contribute to the sustainability and expand the positive externalities.

Data Collection Methods

From the data collection perspective, multiple IoT devices were deployed by all case companies. Many researchers have emphasised the benefits of combining IoT devices and blockchain (Bermeo-Almeida et al., 2018; Van Hoek, 2019). Van Hoek (2019) also suggested that blockchain and IoT devices are complementary, rather than substitutes for each other. RFID was chosen by BeefLedger after it compared multiple IoT devices. RFID can be used to increase the data accuracy and data collection efficiency. In many places, data are recorded manually, which can increase the chance of mistakes and slow down the process. Different from BeefLedger, the QR code was used to collect data for suppliers for W company. The main reason is that the QR code is much cheaper and more affordable for suppliers compared to other IoT devices. W company has thousands of suppliers, many of whom supply food products with little added value, such as vegetables. Therefore, it is important to make sure that suppliers can afford the costs. By using the QR code, the only investments for suppliers are code scanners which only cost a few hundred yuan, or a smart phone with an APP that can scan the code.

FairChain adopted a SmartFarm APP for data collection. The APP is designed specifically for plantation crops and can digitise the on-field practices. SmartFarm APP replaced the old way of paper and pen data collection. Via the APP, FairChain can simply collect data with a smart phone and send data directly to the blockchain platform. Moreover, the APP can also provide solutions to farm management, such as weather advisory, output predictability, and generating reports. Due to limited digitisation in rural areas in Ethiopia and Australia, FairChain and BeefLedger combined online and offline data-uploading methods to ensure data can still be collected without network connectivity. Once there is network access, data that are collected offline will be uploaded on blockchain automatically. This method is particularly useful for areas with poor unstable network.

7.1.3.2 Supplier Engagement

Blockchain is only an infrastructure and it requires all key stakeholders on board. One of the most important things during blockchain implementation is to persuade stakeholders and gain their interest. At this stage, only the key stakeholders who are directly involved in the supply chain were contacted. Thus, suppliers are the key enablers to implement blockchain and participate in the project.

All three companies had provided sufficient education and training to suppliers and presented them the benefits and potentials of blockchain. In order to do so, BeefLedger employed professionals to communicate with suppliers and check their attitude to innovations. The education from a market perspective has also successfully gained suppliers' interests.

Similarly, W company held a supplier conference to discuss blockchain. Food industry consultant—PwC was invited to give their advice and insights about blockchain and its future. The suppliers that had closer relationships with W company were contacted to be on the chain first. Moreover, suppliers who are willing to participate in the project can list their blockchain-traceable products in a better shelf place. Suppliers will also receive sufficient and dedicated training from PwC about how to use the blockchain system. In addition, for suppliers, there are little investment as they are not required to cover the blockchain building costs nor need to invest on new facilities. The only costs for some large suppliers are code scanners. For small suppliers, there is no extra cost on scanner. A smart phone can also do the job. Therefore, suppliers have no financial pressure and so are more motivated to try new technology. In addition, suppliers are concerned about the impact of the blockchain system on product efficiency. VeChain has designed the system to be easily used by suppliers. There have been no major changes to suppliers' operations, but the system can provide obvious benefits.

FairChain is dealing with suppliers who have different culture and language. The first thing for FairChain is to build a relationship with farmers and gain their trust. To attract farmers, FairChain also increased the purchasing price by 20% more than the market price. FairChain was supported by local communities and NGOs in farmer education and communication. The help from NGOs greatly shortened the distance between FairChain and farmers. It has laid a solid foundation for farmers to accept blockchain. From training, farmers can gain the knowledge about the technology and understand what it can do for them. The results of the digital literacy research by FairChain suggests that most farmers, particularly younger generations, have more interests in technology, growing awareness of the blockchain values, and strong urges to improve their income. Therefore, in the FairChain case, farmers are the active participants in the blockchain project.

7.1.4 Inter-Organisational Learning

7.1.4.1 Stakeholder Collaboration

Consumers' Engagement

After blockchain implementation and persuading suppliers, the next step is to invite all the stakeholders including consumers and third parties to share the value of blockchain. From a consumer perspective, the three case companies have different demands and expectations. For BeefLedger and W company, they hope to receive feedback from consumers and collect consumer information. The consumer opinions can benefit both focal companies and suppliers. For example, they can get to know how consumers feel like about the products, and how much confidence can blockchain bring to them. These feedback and information can eventually benefit the sales. However, consumers can only access the data on blockchain but cannot give feedback to the products yet. BeefLedger and W company hope to enable consumer feedback module at a later stage.

One of the restrictions for consumer data collection is the data protection law. To solve the issue, both BeefLedger and W company have a legal department for law and policy-related issues. Another issue for consumer engagement is the cultural difference for the cross-border supply chain. W company and FairChain have no such issue because their targeting groups are mainly local consumers. Therefore, a unique barrier for BeefLedger to collaborate with the consumers is the difference in policy and culture. The cultural difference requires BeefLedger to understand more about the Chinese consumers' living condition, habits, and value perceptions. BeefLedger has held a few roadshow cases in China and has conducted research about consumers' value perceptions. Moreover, BeefLedger also collaborated with Chinese e-commerce—LibertyPost and local agents to sell the products by the ways of live streaming and community buying. The political difference is shown in the China firewall, which restricted Chinese consumers from accessing the blockchain directly. Therefore, BeefLedger has created a server on the China side and moved the data from Australia to China.

However, FairChain has done it differently. FairChain encourages consumers to contribute to the shared value chain by giving tokens to consumers. Consumers can receive and activate a token after each purchase. The token can redeem a discount for the next order, or it can be used to tip the farmers and create a closed value loop. Therefore, consumers can be motivated to become the impact consortia and enhance the positive externalities. Moreover, FairChain is also planning a Microloan Project to enhance the consumer participation experience. By lending farmers more tokens, consumers can help farmers to grow their businesses and create an economic cycle of growth.

Suppliers' Engagement

At this stage, companies are engaging more suppliers on board, including more upstream suppliers. In other words, as blockchain has been implemented and accepted by some key suppliers, companies can expand the scale of suppliers on the chain. For BeefLedger, upstream suppliers mean the processes before producers, such as feedlot. For W companies, only direct suppliers from the factory side were first invited. Thus, they are expanding the traceable supply chain to the raw material side in this step. For example, rice products can be traced back to the plantation process. Horizontally, W company also invites more suppliers from the international level.

Different from BeefLedger and W company that engage upstream suppliers until the late stages of the project, FairChain engaged upstream suppliers—farmers at the very beginning. With the progression of the blockchain project, FairChain has started to involve higher-tier suppliers, including coffee graders and processors. The more suppliers on the blockchain, the more transparent the economic model and the cost of each stage become.

Third-Party Support

From a third-party perspective, all three companies received different levels of support from the governments and NGOs. BeefLedger and FairChain successfully applied for government funding on the projects. W company also claimed that the Chinese government has been supportive on the project, particularly during the pandemic. The government encourages food companies to explore the potentials of blockchain on ensuring food safety. NGOs are specifically helpful in the case of FairChain. FairChain shares different culture and language to the coffee farmers, so NGOs and local communities are playing the role of the bridge that connects FairChain and farmers. They not only communicate with farmers, but also responsible for farmer education. Therefore, although they are not directly involved in the blockchain projects in all three cases, governments and NGOs are necessary for facilitating the blockchain implementation and are the influencing factors for the success of the projects. FairChain is also planning to invite auditing companies to participate in the project in the future to enhance the data credibility.

7.1.4.2 System Adjusting and Business Adapting

This is the step that follows from of the blockchain implementation and stakeholder collaboration. The new insights and understandings that emerge during implementation and can lead to the system needing updating and business adapting. The changes have two perspectives: technology perspective and business perspective.

Technology Perspective

All the three case companies are planning to improve the user experience from the supplier side and consumer side, and to make the system more user friendly. Among all the aspects that influence user experience, interface design is one of the key features. BeefLedger claimed that they are trying to build a system that is as easy to use as possible. Moreover, they are also working on knowledge management. Rather than giving every detail to consumers, they classify different information and let consumers choose what they want to know. This can largely avoid information overload and feeling of fatigue.

W company also agreed on the importance of user experience. From the supplier side, VeChain had redesigned the system and added new functions to make the system become more user friendly. For suppliers, one of the requirements for the blockchain system was that it should not affect the existing operations. W company had optimised the system to cope with suppliers' requirements. For consumers, the information pages were designed to be more interesting, for example, the background colour matches the product features. From the consumer side, at the very beginning of the platform building, VeChain had the idea of designing a dedicated APP for consumers to download to check the product information. However, this can increase the barrier level and discourage consumers from using it. Therefore, to bring convenience and motivate consumers, W company decided to use an attached QR code on the product package. Consumers with a smart phone can check the product information. Meanwhile, to prevent QR code counterfeiting, W company allocated a QR code per item instead of per batch, which leaves no gaps or profits for counterfeiters.

FairChain has conducted semi-structured interviews about local digital literacy at the data collection stage. The research suggested that farmers were suffering from high illiteracy, particularly for people aged over 40. Thus, FairChain was constantly changing the system design to be more farmer friendly. For example, the system was designed with more pictures and a listening function for farmers who cannot read. In addition, FairChain also saw the young generation as an opportunity to promote the new technologies. The younger generation showed more passion for and curiosity about the new technology and were more eager to learn. Therefore, they were the great driving force to the adoption of blockchain.

Apart from interface design, all three companies are planning to add more functions to the blockchain system. BeefLedger has plans to enable smart contract soon to smooth payment and documentation. As mentioned above, time is particularly important for the cross-border beef supply chain. The duration is the key influencing factor for product turnover and operation costs. Therefore, enabling smart contract can reduce a significant amount of processing time in document checking. Payment can also be generated automatically to reduce hassle. To smooth the payment process, W company also has a plan to use smart contracts in the future. FairChain is planning to move from a private to a public blockchain. Currently, the private blockchain only allows certain users to access the chain. After moving to a public blockchain, the transparency can be enhanced as the data can be more accessible for the public.

Business Perspective

From the business perspective, all three companies have plans on expanding their operations. BeefLedger is providing the blockchain solutions to other food brands. Supply chain finance is also another direction that BeefLedger hopes to explore in the future. W company is involving more products and suppliers on the chain. W company is also adjusting the targeting group and changing it from supercentres to Sam's club—a middle- to high-end store. Consumers in Sam's club tend to show more interest in food quality and safety. They have a stronger need to know more about product information. FairChain is expanding the product scales on the chain such as chocolates. Not only for products; the sustainable practices can also be traceable on the chain. FairChain is operating different sustainable solutions such as Microloan and coffee waste recycling. Therefore, the companies continue to be flexible and open minded to update blockchain to fit their business needs.

7.2 Critical Success Factors

This section proposes four factors (capabilities, collaboration, technology readiness, external environment) and eight sub-factors based on the observations on the critical success factors during blockchain implementation (Fig. 7.2; Table 7.2). Table 7.2 summarises these factors and provides the supporting quotations.

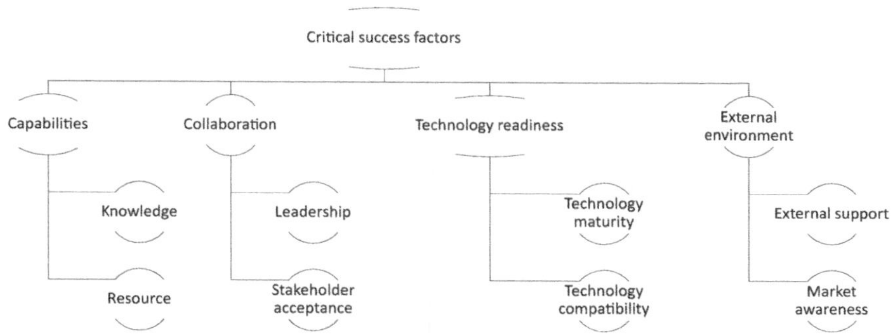

Fig. 7.2 Critical success factors' structure (Source: the author)

7.2.1 *Capabilities*

The first key success factor is a company's capabilities, which normally mean a company's ability to adopt an innovation. Two sub-factors are knowledge and resources.

7.2.1.1 Knowledge

To gain sufficient knowledge, the three case companies adopted different approaches. BeefLedger and FairChain conducted several research studies related to blockchain and the food industry. W company proposed various business plans to discuss with its key business partners. Besides, all three companies had pilot runs to gain practical experience before officially launching their blockchain projects.

7.2.1.2 Resources

From a financial perspective, as start-ups, BeefLedger and FairChain received funding from governments. As one of the global leading retailers, W company has fewer concerns about financial pressure. Therefore, all case companies have sufficient financial resources to implement blockchain. From a human resource perspective, every case company received support from professionals, including technology and industry support.

7.2.2 *Collaboration*

The second success factor is the collaboration between stakeholders during blockchain implementation, including the lead company's leadership and other

Table 7.2 Critical success factors

Critical success Factors	Sub factors	BeefLedger	W Company	FairChain
Capabilities	Knowledge	1. BeefLedger conducted a year-long-research study about blockchain technology. 2. BeefLedger ran a pilot to trace products from Australia to China. "We did about a year of internal research and development before launching the project. So we officially launched the project around October, November, 2017. So, we began our internal research and development and business case mapping in 2016".—Warwick Powell, Chairman of BeefLedger	1. W company has announced two pilots since 2016, so has gained rich practical experience. 2. W company discussed several business plans with business partners—PwC and VeChain. "W company, American branch and Chinese branch had already tried out some POC with IBM on blockchain from 2016 to 2017".—Mr. Feng, senior manager at PwC food & beverage retail. "I have always been in the food industry, and I have known some of their managers for nearly 10 years. We first talked about these concepts (blockchain) in an industry conference. Turns out that they were very interested. Therefore, we kept discussing it and trying to find its the commercial values and proposing and verifying business plans for over half a year. Then, we started to set up the project".—Mr. Feng, senior	1. FairChain conducted a pilot before officially implementing blockchain. 2. FairChain has conducted a few research studies, including interviews, to understand farmers' living conditions and to design the "income road map" to help farmers improve their living incomes. "The first mile means that the farmer has a face. It's not just a farmer, but we know all our farmers by name, we have their faces and their profiles. We know what they're producing, how big their families are. And now we're capturing data that will contribute to our long-term ambition to bring them to a living income level".—Mr. R, the chief technology officer of FairChain Foundation.

(continued)

Table 7.2 (continued)

Critical success Factors	Sub factors	BeefLedger	W Company	FairChain
			manager at PwC food & beverage retail.	
	Resources	1. BeefLedger partnered with Food Agility and QUT to conduct research about blockchain and to run pilots. 2. The founders of BeefLedger have been working in the food industry for a long time and have gained rich experience. "BeefLedger is a technology start-up company, which means that they had a lot of resources in the industry in the past. They are very familiar with the Australian beef industry, and then they discovered these problems and opportunities in the industry. So, they are setting up a company, and developing blockchain application in the industry".—Dr. Cao, BeefLedger researcher "BeefLedger cooperated with QOT and another Australian funding organisation—food agility, on the project".—Dr.	1. W company invited food industry expert—PwC and blockchain professional—VeChain to jointly discuss and built the blockchain platform. 2. W company is one of the largest retailers in the world and has strong financial ability to invest in a new technology. "It is not purely about technology; the food industry is complicated. So, during the implementation, if you do not understand this industry, it is going to be very difficult. PwC is here because we have a dedicated food team, we are the only one in the Big Four that has a consulting team specialising in the food industry. Combining our industry and consulting knowledge, and we have a close business relationship partner—VeChain. All those factors made W company decide to work with us".—Mr. Feng, senior	1. FairChain partnered with KrypC—a blockchain expert, to jointly build a blockchain platform. 2. FairChain applied for government funding to support the blockchain project. "I source it via a company called KripC, and they have extensive knowledge on Hyperledger and now to build blockchains on that. And together we design blockchain. So, we do the service design together and they build the intellectual properties for FairChain".—Mr. R, the chief technology officer of FairChain Foundation. "So the cost of for development of the platform is partially private money and partially governmental funding from the Dutch government. For running the systems, that's what we have a

(continued)

Table 7.2 (continued)

Critical success Factors	Sub factors	BeefLedger	W Company	FairChain
		Cao, BeefLedger researcher	manager at PwC food & beverage retail.	FairChain entity for, we sell software solutions to the market on a cost-plus model. So I have a list of applications we have and the list of tools we can provide to the market and our consumer".—Mr. R, the chief technology officer of FairChain Foundation.
Collaboration	Leadership	1. BeefLedger lead by example to motivate stakeholders. 2. BeefLedger is helping to improve local digitalisation where necessary. "Let's say we use our beef supply chain as an example to let others find that blockchain is indeed useful. They can get more consumer information and understand the consumer's behaviour to promote their products. We hope they can see the value and will join us gradually"—.Dr. Cao, BeefLedger researcher	1. W company has covered all the blockchain building and developing fees to reduce suppliers' financial pressure. "Because W company wanted to make a faster move, so they decided to cover all the costs, including supplier training and consulting costs".—Mr. Feng, senior manager at PwC food & beverage retail.	1. FairChain first tried to build a close relationship with farmers to gain their trust. Therefore, FairChain not only paid higher than market price for purchasing coffee cherries, but also financed a wet mill to provide local more job opportunities. "For farmers there is a big incentive. They get higher prices because the business model of FairChain is based on higher prices for farmers. There's long-term benefit as consumers might plant the tree and farmers one day wake up and receive trees from consumers".—Mr. A, the

(continued)

Table 7.2 (continued)

Critical success Factors	Sub factors	BeefLedger	W Company	FairChain
				FairChain Architect.
	Stakeholder acceptance	1. BeefLedger employed professionals to check suppliers' acceptance of new technology. 2. BeefLedger gave suppliers freedom to choose what information can be uploaded on the chain. "We have a professional who is checking producers' acceptance from a cultural perspective, and then educate the producer through the market perspective. For example, by discovering consumer data and feedback, that means that after the producer enters our ecosystem, they can view consumer feedback and understand the dynamics of your product. In addition, if some companies think this thing is great, they may recommend some other people to use it through the recommendation system".—Dr. Cao, BeefLedger researcher "If you want everything on the chain, you have to	1. W company invited the Tier 1 suppliers who built a close relationship with W company. Some of the suppliers are W company's own brand suppliers. 2. W company provided incentives to attract suppliers to participate in the project. For example, traceable products can be listed in a better shelf place. 3. W company arranged supplier meetings to discuss blockchain and to educate suppliers. PwC will give full training for suppliers who agree to join in. 4. W company selected QR code instead of more expensive devices to encourage suppliers to participate. "We only need to persuade W company to launch the project. As soon as they approve the project, they communicate with their suppliers, particularly the suppliers who have good relationships with them. They	1. FairChain not only raised purchasing price to increase farmers' income, but also financed a wet mill to provide farmers more jobs. 2. FairChain arranged various training on agriculture, technology, and finance to educate farmers. 3. FairChain has looked for companies that share the same sustainability goal with them. "So in the beginning you had to find farmers who want to supply cherries to you. Of course, paying a better price always helps. But they're now just lining up, they want to be part of this FairChain farmer programme".—Mr. R, the chief technology officer of FairChain Foundation. "And so we bought this this wet mill for the community. And so basically we financed it. We bought it and over time the farmers and the

(continued)

Table 7.2 (continued)

Critical success Factors	Sub factors	BeefLedger	W Company	FairChain
		have an on-chain and off-chain governance structure. This involves a problem of governance structure on the chain. For example, there may be some information that companies are willing to publish, and some information that companies are unwilling to publish, that is, they can only say. Just like on our platform, we have also designed a group's permissions. As a participant, you can set which information you want publish to the entire chain for others to see, and which may even have many protections. You can say which information you can let upstream people see, which information can be seen by people in the next step, and which information can only be seen by some executives of the Company".—Dr. Cao, BeefLedger researcher	persuade them from the perspectives of food safety and marketing".—Mr. Cheng, the project manager in VeChain "In this project, system building is not very difficult, the difficulty is how can you persuade suppliers to join in? Some companies might just want to have a blockchain-based internal information system, and they can simply require all departments to use it. However, for W company, the platform participants are not directly under its control. Moreover, the precondition for many suppliers to adopt the system is that the system should negatively impact their production efficiency and logistics. That is why we say the key things we need to consider during system implementation are the system efficiency and the sustainability of the operation".—Mr. Cheng, the project manager in VeChain "Suppliers need to	stakeholders in the FairChain, they become owners of this wet mill as well. We leave value-added activities in Ethiopia, like roasting and packing, etc. so that creates jobs and that creates income at the end of the line".—Mr. R, the chief technology officer of FairChain Foundation. "It's also a part of a selection of partners. So, we say, we want to be transparent about the cost and the things we do. And we select companies that also want to comply with that, so that it's not that difficult. Our objective is to make sure everybody makes a proper income out of our value chain and people understand that. So, we don't need to find people, they are knocking on our door".—Mr. R, the chief technology officer of FairChain Foundation.

(continued)

Table 7.2 (continued)

Critical success Factors	Sub factors	BeefLedger	W Company	FairChain
			use our APP, and need to log in to the backend for information uploading. Our team needs to teach suppliers how to use the system and how to upload information, both online and on-site".—Mr. Feng, senior manager at PwC food & beverage retail. "We have done a very accurate cost calculation. The suppliers only need to buy a code scanner per production line. The scanner can be as cheap as from 200 Chinese yuan. For smaller suppliers, if the production is not much, the smart phone with an APP can also do the job".—Mr. Feng, senior manager at PwC food & beverage retail.	
Technological readiness	Technology maturity	1. BeefLedger is trying new infrastructure to solve the scalability issue. 2. BeefLedger has enabled multi-signature function to verify data. 3. RFID devices are used to collect data. "The first category of challenges	1. The majority of the information is saved off-chain. 2. W company can temporarily hide wrong information from consumers. 3. Data are checked by third parties. 4. QR code is adopted for data collection. "We did not store all the information	1. FairChain holds a positive attitude to scalability issues. 2. Triple data checks are made before data uploading. 3. IoT devices are used to collect data. "I don't see scalability is a problem. Scalability is

(continued)

Table 7.2 (continued)

Critical success Factors	Sub factors	BeefLedger	W Company	FairChain
		relates to the technology itself. The technology itself is a young technology. So, there are always issues in terms of the building of complex systems with young technologies".—Warwick Powell, Chairman of BeefLedger "In fact, the challenge we are facing is the scale of the Ethereum platform we use now; how can we extend its scalability to carry large transaction volumes in the future? Its infrastructure is in the process of slow development and has not yet achieved such a strong transaction speed or carrying capacity. Some people are not talking about using cloud storage. You saw it consumers can view not just text, but also some short videos. For things like this, if you produce every transaction, in the end, it means that the data storage is huge at that time. One of our staff has mentioned Holo chain. Then it	on chain because it will slow it down. While a consumer is scanning the QR code, there are always a few seconds verification process with the blockchain background. If the code is incorrect, the information will fail to display. Once the verification succeed, consumers not only can see the encrypted product information, but also can also see a blockchain transaction hash (TXID), which can be verified on VeChain's public blockchain".—Mr. Feng, senior manager at PwC food & beverage retail. "Of course, to prevent some mistakes that are maybe complained about by consumers, typos such as grams instead of kilograms, the one thing we can do is to hide the wrong information. Instead, we link the traceability system to the official website".—Mr. Feng, senior manager at PwC food & beverage retail. "It is more	really . . . or replicability is now what we call it. Scalability is always a short-term problem. It's literally the same thing as in 10 years, 15 years ago you couldn't attach 500 kilobytes to an email and if you want to upload a photo, then it was three MB and you couldn't do that. So, I think innovation will always lead before scalability".—Mr. A, the FairChain Architect. "We have the local team on the ground who does the job, they check their own job. Moyee does a coffee as our partner in this coffee collection; they are the contacting party with the farmers, and they check that what the local team does is right. And FairChain is the third and final step after there's sort of a triple check on whether the data is correct. If all of the transactions are nicely made before we publish

(continued)

Table 7.2 (continued)

Critical success Factors	Sub factors	BeefLedger	W Company	FairChain
		means that some of these problems may be solved by Holo chain. At that time, he said that it is equivalent to establishing a regional database, we also want to use the BSN in China, to link or upload a part of data on there".—Dr. Cao, BeefLedger researcher	effective to formulate the blacklist system, so once the data has been uploaded, you cannot change it, you will have to be responsible for it. We (PwC) as a third party to check data authenticity, W company might also do it themselves".—Mr. Feng, senior manager at PwC food & beverage retail.	them on the blockchain, that's why we use this triple accounting. We validate if we got proof for harvest collection, we validate with the payments slip, and with the bank that the money is actually being transferred. You know, I have triple account. There's also a triple accounting mechanism in blockchain. We rely on the multiple sets of eyes checking data before it's entered in the bucket".—Mr. R, the chief technology officer of FairChain Foundation. "The big solution to this is called IOT, Internet of Things. So the smart scales and smart sensors lead the way to prevent intervention or the human touch towards entering data".—Mr. A, the FairChain Architect.
	Technology compatibility	1. BeefLedger focused on the blockchain design to be user friendly. 2. BeefLedger decided to use RFID to accurately	1. W company designed the blockchain system to be user friendly for suppliers. 2. W company selected QR code	1. FairChain adopted SmartFarm App for data collection to collaborate with blockchain. "I think that

(continued)

Table 7.2 (continued)

Critical success Factors	Sub factors	BeefLedger	W Company	FairChain
		collect data. "One of the challenges is being able to make this technology as easy to use for people as possible. And the technology, the blockchain industry itself I think actually understands that's probably one of its weakest parts; that it is still not as easy to use and as easy to interface with as more conventional information systems, but this is improving".—Warwick Powell, Chairman of BeefLedger "But when you are using blockchain technologies as a ledger structure to interface with the real world, you need to have ways in which the blockchain architecture relates to the external environment. And so you need to have different kinds of informational interfaces at the production or at the collection. And so Internet of Things, data is one, but basically you just need computers".—Warwick Powell,	to be the data collection method to reduce supplier financial pressure and to encourage suppliers to participate in the project. "Why we say that the key to operate and apply blockchain application depends on the construction of the entire system, and if the system has a minimum impact on suppliers' production efficiency? If the impact is little, they are usually happy to cooperate with W company to do things. This is also an opportunity to market their products at the same time".—Mr. Cheng, the project manager in VeChain "With QR code, it costs a few cents per label. You still need to consider the costs. For example, five cents are 1% of the gross profits for a box of vegetables that cost five Yuan (1 Yuan = 100 cents). If we use RFID, the costs are at least ten times of QR codes".—Mr. Feng, senior	blockchain is not the Holy Grail, but the combination of blockchain with those IOT machines and machine learning and AI systems are bringing the whole package towards the level that it needs to be. So, it's not a substitute for blockchain. I think blockchain is one of the key components because it's just a ledger that is distributed in which everybody can just look at the same information. And I think that is absolutely key to have. I think the infrastructures of blockchain will have to improve and also in regards to privacy. But I think that there is not a substitute for blockchain, but I think that only having blockchain is not good enough. It doesn't cover everything you need to work with in collaborating and bringing other technologies in; it is only very distinct by name. But on a tech level, it is very much

(continued)

Table 7.2 (continued)

Critical success Factors	Sub factors	BeefLedger	W Company	FairChain
		Chairman of BeefLedger "We work with device manufacturers who specialise in temperature gauges, who specialise in geospatial tracking. They specialise in asset identification such as RFID asset identities who provide technology that enables us to collect information from vehicles. We work with people who have developed devices that are attached to the animals to measure our animal temperatures and animal heart rates and blood oxygen levels so that we could monitor animal health and conditions. So we do need to, and we do work with people who are involved in many specialised areas".—Warwick Powell, Chairman of BeefLedger	manager at PwC food & beverage retail.	connected".—Mr. A, the FairChain Architect.
	External support	1. BeefLedger received government funding to support running the project. "In fact, our project is a business plus university plus industry, and then project funds come	1. Chinese government shows a supportive attitude to the project. "The government actually pays attention to it. At the national level, companies are encouraged to do	1. FairChain applied for government funding to support the project. 2. FairChain invited NGOs and local communities to communicate with and provide

(continued)

Table 7.2 (continued)

Critical success Factors	Sub factors	BeefLedger	W Company	FairChain
		in. For example, if we do this project, because this project is an Australian funding organisation, they also donate part of the funds, and then BeefLedger also provides part of the funds, and then it is equivalent to a project with a total value. It is not an operation of a pure enterprise, it is a kind of cooperation that is equivalent to enterprise + industry + funding organisation, which means co-promotion".—Dr. Cao, BeefLedger researcher	such things (blockchain-enabled traceability), because it is meaningful for people's livelihood".—Mr. Cheng, the project manager in VeChain	training to farmers. "Governments are always behind in innovation. That's not the problem. Innovation happens. And governments see it happening. They see it failing, they see it flourishing, and they provide the regulatory frameworks to scale it and make it responsible".—Mr. A, the FairChain Architect. "We work with the universities and some of those kinds of NGOs. We work with the Dutch government, Ethiopia, the Dutch embassy in Ethiopia. So, there are more people on that".—Mr. R, the chief technology officer of FairChain Foundation.
	Market awareness	1. Consumers have requirements on the information of food products. 2. Companies in the food industry need to meet consumers' satisfaction in order to stay competitive. "Usually we find particularly in the	1. The new generations are changing their consumption habits from brand loyalty to caring more about product information. "I think this seems to be an extra layer of protection for the consumers. Anyway, a layer	1. Consumers gradually gain awareness of the shared value chain. "By starting with food is because it's the most accessible for consumers. The only thing that you have to do is to

(continued)

Table 7.2 (continued)

Critical success Factors	Sub factors	BeefLedger	W Company	FairChain
		China context, their first is validation of country of origin stuff that I just want to know where it came from. And that's the first thing. The second thing that matters is that the product is safe, which means that the basic measure of that is the cold chain logistics information on temperature control".—Warwick Powell, Chairman of BeefLedger "The market will move quicker than that. Social networks and social eCommerce and social media make the markets work a lot quicker than they ever did. So consumers will move incredibly fast, in the face of getting something for nothing. It's about sustaining a competitive position because in the end, if other businesses don't do it, they will be the ones who get discounted".—Warwick Powell, Chairman of BeefLedger	more is better than nothing".—Mr. Feng, senior manager at PwC food & beverage retail. "To be frank, people value differently. For older generations, they believe in more of the brand. For some millennials, there are changes in their consumption habits. Their loyalty to the traditional brand will be lower and lower. They are easy to be attracted by something new, such as the sustainable packaging or food safety, etc. they are more likely to be different to old generations … After the consumption habits change, everyone keeps advocating and everyone will scan it. After getting used to it, you will actually check it every time you buy it. You do not only believe the shelf life printed above, but trace the information displayed in the QR code, whether this thing is indeed qualified, safe, and reliable".—Mr. Cheng, the project manager in VeChain	create a better life for a farmer by buying a coffee, which you are most likely doing several times a week in your natural behaviour. You don't buy a luxury watch a couple of times a week. You do buy coffee a couple of times a week. So our model is based on changing existing consumer behaviour, or not changing it, leaving consumer behaviour as it is, and then providing that with another step in which they can make more impact".—Mr. A, the FairChain Architect.

stakeholders' acceptance of the blockchain. Two sub-factors are leadership and stakeholder acceptance.

7.2.2.1 Leadership

The three companies used a range of methods to motivate suppliers. BeefLedger helped their suppliers to solve problems related to low digitalisation. W company helped suppliers to solve financial concerns by covering all the implementing costs. FairChain built a close relationship with the farmers by purchasing cherries at high prices and financed a wet mill to provide more jobs to the local area.

7.2.2.2 Stakeholder Acceptance

Stakeholder acceptance refers to if stakeholders are willing to adopt blockchain. To encourage suppliers, BeefLedger employed specialists to talk to suppliers about blockchain and gave suppliers freedom to upload information that they are willing to share. W company provided incentives to suppliers by listing blockchain-traceable products in better place and collaborated with a food expert—PwC to give sufficient training. Moreover, considering suppliers' financial pressure, W company adopted the QR code instead of others to reduce costs and encourage suppliers to participate. FairChain also first provided training to farmers and increased their understanding of blockchain. Next, FairChain invited stakeholders who share the same goal of promoting sustainability to jointly build a blockchain platform.

7.2.3 Technology Readiness

Technology readiness refers to blockchain maturity to adopt and if blockchain is compatible to the company's current system. Two sub-factors are technology maturity and technology compatibility.

7.2.3.1 Technology Maturity

Blockchain is a young technology, so there are a few issues relating to technology maturity can influence a company's decision-making. One of the technology-related issues is scalability, BeefLedger is trying new infrastructure to allow scalability issues to be solved. W company combined on-chain and off-chain methods to store information. FairChain has no concerns about scalability. Raw data mistakes are also a side effect due to technical restraints. All the companies adopted IoT devices to collect data and reduce data mistakes. BeefLedger and FairChain also enabled multi-signature function and triple data checks to enhance raw data authenticity.

7.2.3.2 Technology Compatibility

To improve technology compatibility, all companies designed the blockchain platform to be more user friendly and adopted cheap IoT devices for data collection. Thus, suppliers are not required to replace the whole system by blockchain but combine blockchain with the current system. It, therefore, can reduce the financial pressure of suppliers and make blockchain implementation more convenient.

7.2.4 External Environment

External environment includes encouragement from third parties, and the market's needs from and reactions to blockchain. Two sub-factors are external support and market awareness.

7.2.4.1 External Support

As mentioned above, BeefLedger and FairChain received government funding to support the blockchain projects. This support from third parties can also influence the blockchain implementation. For W company, Chinese government shows a positive attitude to encourage blockchain implementation in food supply chains. Apart from government, FairChain is also supported by NGOs and local communities on communicating with farmers.

7.2.4.2 Market Awareness

The market reaction to blockchain can also influence a company's decision-making. More consumers have higher requirements of supply chain transparency and tend to know more about the food products. Therefore, case companies are using blockchain to meet consumer expectations by providing more information to consumers.

7.3 Common Barriers and Solutions

This section summarises the common barriers during blockchain implementation, and how companies overcame the barriers (Fig. 7.3, Table 7.3). The barriers are divided into four categories based on the framework that was suggested by Saberi

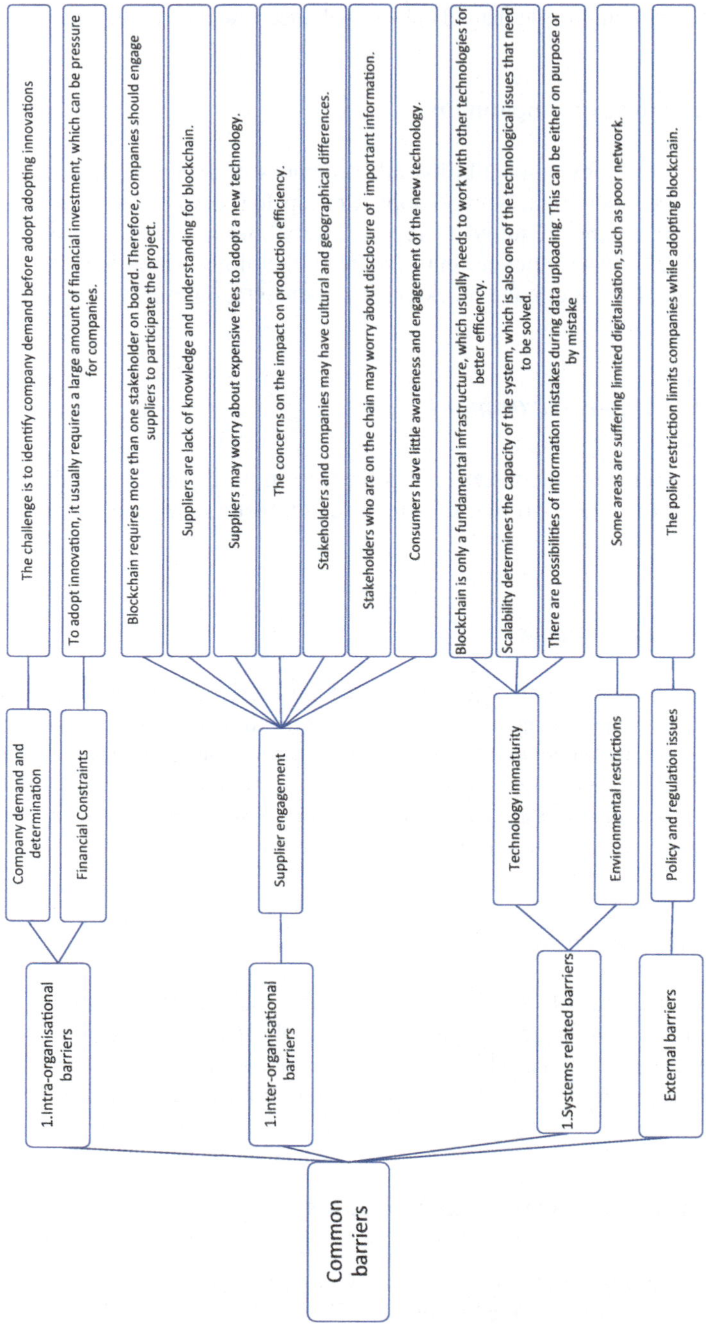

Fig. 7.3 Common barriers of the three case companies

Table 7.3 Common barriers and company solutions

Aggregate construct	2nd construct	1st construct	Solutions and supporting quotations
Intra-organisational barriers	Company demand and determination	The challenge is to identify company demand before adopting innovations.	1. All three case companies had not only invited experts to discuss the project, but also conducted several research studies to gain a deep understanding of the technology in order to find the best fit. "We did about a year of internal research and development before launching the project. So, we officially launched the project around October, November, 2017. So we began our internal research and development and business case mapping in 2016".—Warwick Powell, Chairman of BeefLedger
	Financial constraints	To adopt innovation, it usually requires a large amount of financial investment, which can be pressure for companies.	1. Apart from W Company which is a world-leading retailer with strong financial background, BeefLedger and FairChain, as SMEs, received funding from governments. "In fact, our project is a business plus university plus industry, and then project funds come in. For example, if we do this project, because this project is an Australian funding organisation, they also donate part of the funds, and then BeefLedger also provides part of the funds, and then It is equivalent to a project with a total value It is not an operation of a pure enterprise, it is a kind of cooperation that is equivalent to enterprise + industry + funding organization, which means co-promotion".—Dr. Cao, BeefLedger researcher
Inter-organisational barriers	Stakeholder engagement	Blockchain requires more than one stakeholder on board. Therefore, companies	1. Companies invited suppliers who have close relationships with them.

(continued)

Table 7.3 (continued)

Aggregate construct	2nd construct	1st construct	Solutions and supporting quotations
		should engage suppliers to participate in the project.	"We only need to persuade W company to launch the project. As soon as they approve the project, they communicate with their suppliers, particularly the suppliers who have good relationships with them. They persuade them from the perspectives of food safety and marketing".—Mr. Cheng, the project manager in VeChain 2. Companies have provided incentives to attract suppliers. "People think that blockchain technology is simply about software and coding, but at the heart of blockchain is the challenge for people in supply chains, for example, to find ways of collaborating. So, it is the technology that requires collaboration. So, the challenges are not about technology, they're about collaborative culture and the issues involved in achieving collective action around things that people have common interests about".—Warwick Powell, Chairman of BeefLedger
		Suppliers have lack of knowledge and understanding for blockchain.	1. Sufficient training was provided to suppliers. "In this project, system building is not very difficult; the difficulty is how can you persuade suppliers to join in. Some companies might just want to have a blockchain-based internal information system, and they can simply require all departments to use it. However, for W company, the platform

(continued)

Table 7.3 (continued)

Aggregate construct	2nd construct	1st construct	Solutions and supporting quotations
			participants are not directly under its control. Moreover, the precondition for many suppliers to adopt the system is that the system should negatively impact their production efficiency and logistics. That is why we say the key things we need to consider during system implementation are the system efficiency and the sustainability of the operation".—Mr. Cheng, the project manager in VeChain
		Suppliers may worry about expensive fees to adopt a new technology.	1. Focal companies covered all blockchain platform building fees. Suppliers are not required to invest much money to participate in the project. "Because W company want to make a faster move, they decided to cover all the costs, including supplier training and consulting costs".—Mr. Feng, senior manager at PwC food & beverage retail.
		The concerns on the impact on production efficiency.	1. The blockchain platform was built to be supplier user friendly. "The biggest problem we encountered during implementation is how to improve the efficiency. In this area, for example the information input, they have repetitive tasks every day. We had to consider how to reduce the repetition and the workload. That's why we say that the key to operate and apply blockchain application depends on the construction of the entire system, and if the system has a minimum impact on suppliers' production efficiency. If you

(continued)

Table 7.3 (continued)

Aggregate construct	2nd construct	1st construct	Solutions and supporting quotations
			need a dedicated staff to spend a long time on the system; plus many errors can be easily made during data input which leads to rework. If this is the case, suppliers will have a lot of concerns about using the system. If the impact is little, they are usually happy to cooperate with W company to do things. This is also an opportunity to market their products at the same time".—Mr. Cheng, the project manager in VeChain
		Stakeholders and companies may have cultural and geographical differences.	1. Companies conducted research about consumer behaviours and values. "In fact, the thinking modes of Chinese consumers and the Western consumers are completely different. For example, if you design a kind of social software or something similar based on the consumption mode and consumption behaviour of the Westerns, the Chinese consumers will feel like it is impossible to use it. But if you take it to the Western world, it will be accepted. If you get something that is very good in Australia and bring it to China, Chinese consumers may not want to use it".—Dr. Cao, BeefLedger researcher 2. Local teams were invited to facilitate the project. "We don't deploy those services ourselves. We work with NGOs that deliver such training. We train the NGOs in that sense. Because it is not a one-time practice. It's an ongoing process all year around. So we cannot

(continued)

Table 7.3 (continued)

Aggregate construct	2nd construct	1st construct	Solutions and supporting quotations
			control that from an office in Amsterdam. So we worked with NGOs that we train how to use the technology and the education of farmers. And that's going quite okay".—Mr. A, the FairChain Architect.
		Stakeholders who are on the chain may worry about disclosure of important information.	1. The suppliers have the right to decide what information to upload on the chain. "If you want everything on the chain, you have to have an on-chain and off-chain governance structure. This involves a problem of governance structure on the chain. For example, there may be some information that companies are willing to publish, and some information that companies are unwilling to publish; that is, they can only say we. . . Just like on our platform, we have also designed a group's permissions. As a participant, you can set which information you want publish to the entire chain for others to see, and which may even have many protections. You can say which information you can let upstream people see, which information can be seen by people in the next step, and which information can only be seen by some executives of the Company".—Dr. Cao, BeefLedger researcher 2. Suppliers are anonymous on the chain. "The farmer data is not in the blockchain itself. It's all anonymised. So a farmer has just a number in the blockchain it's called a

(continued)

Table 7.3 (continued)

Aggregate construct	2nd construct	1st construct	Solutions and supporting quotations
			farmer number something, and we've got an off-chain database where farmers' profiles are in. And so part of the data is in the blockchain itself for immutability, the part of it is like the personal and privacy and GDPR-related data is in an off-chain database and not accessible for anyone".—Mr. R, the chief technology officer of FairChain Foundation.
		Consumers have little awareness and engagement of the new technology.	1. Companies are giving consumers education about blockchain by publicity. "For Sam's club, its target group is mid-to-high-end consumers. They are not very sensitive to the price, but are more sensitive to the product quality, such as country of origins. So, there is some special publicity in the store to tell consumers that the products are traceable on the platform. The products have traceable labels".—Mr. Cheng, the project manager in VeChain 2. Consumers were encouraged by receiving tokens. "So instead of spending the marketing budget for the brand on Facebook or Snapchat or, you know, an advertisement running online or offline, you put that marketing money into the hands of the consumers, so whenever they purchased a cup of coffee, they received a token".—Mr. A, the FairChain Architect.
Systems-related barriers	Technology immaturity	Blockchain is only a fundamental infrastructure, which usually needs to work with other technologies for better efficiency.	1. Companies adopted various IoT devices to work with blockchain platform for accurate data collection. "We tested that RFID and

(continued)

Table 7.3 (continued)

Aggregate construct	2nd construct	1st construct	Solutions and supporting quotations
			NFC and found NFC couldn't upload a batch of data. Through the test, RFID was chosen to be used to get the batch of data".—Dr. Cao, BeefLedger researcher
		Scalability determines the capacity of the system, which is also one of the technological issues that need to be solved.	1. Change blockchain infrastructure. "If you produce every transaction, in the end, it means that the data storage is huge at that time. One of our staff has mentioned Holo chain. Then it means that some of these problems may be solved by Holo chain. At that time, he said that it is equivalent to establishing a regional database. We also want to use the BSN in China, to link or upload a part of data on there".—Dr. Cao, BeefLedger researcher 2. Combining off-chain and on-chain data. "We did not store all the information on the chain because it will slow it down. While the consumer is scanning the QR code, there are always a few seconds verification process with the blockchain background. If the code is incorrect, the information will fail to display. Once the verification succeeds, consumers not only can see the encrypted product information, but also can also see a blockchain transaction hash (TXID), which can be verified on VeChain's public blockchain".—Mr. Feng, senior manager at PwC food & beverage retail.
		There are possibilities of information mistakes during data uploading. This can be	1. Combining IoT devices to collect data. "You use sensors to obtain

(continued)

Table 7.3 (continued)

Aggregate construct	2nd construct	1st construct	Solutions and supporting quotations
		either on purpose or by mistake.	data, after you obtain it, you need someone to verify it. Retailers want to trace the source of responsibility. For example, to trace who is responsible for the product in each link in the process, if something goes wrong, it can be directly isolated".—Dr. Cao, BeefLedger researcher 2. Enabling multi-signature functions. "We have also implemented a multi signatory process by which data is committed to the blockchain itself, which means that we ensure that participants have to take active responsibility for checking the validity of information that they submit. And that you need to have more than one person agree that the information is valid before it can go onto the blockchain".—Warwick Powell, Chairman of BeefLedger 3. Third parties were invited to verify data. "It is more effective to formulate the blacklist system, so once the data has been uploaded, you cannot change it, you will have to be responsible for it. We (PwC) as a third party check data authenticity. W company might also do it themselves".—Mr. Feng, senior manager at PwC food & beverage retail.
	Environmental restrictions	Some areas are suffering limited digitalisation, such as poor network.	1. Companies helped to build the area gradually. "So in our supply chain, for example, there is limited digitalisation in the first two functions of the information

(continued)

Table 7.3 (continued)

Aggregate construct	2nd construct	1st construct	Solutions and supporting quotations
			system; namely, the collection of information or the production of information. And digitisation is a very big challenge for the industry generally".—Warwick Powell, Chairman of BeefLedger 2. Data collection moved to offline when network connectivity is poor. "You cannot rely on online systems. You have to have a sort of offline/semi-online mechanism in place, and that's what we have for everything. And we can enter data without internet connection. And as soon as the internet is up again, and we synchronise the data and do the checks, checks, and checks, and we send it through to blockchain".—Mr. R, the chief technology officer of FairChain Foundation.
External barriers	Policy and regulation issues	The policy restriction limits companies while adopting blockchain.	1. Companies employed a professional legal department to solve regulation issues. "We have a team that specialises in this area, which involves legislation, legal regulation, and how to operate legally and compliantly. There must be a lot of considerations for this blockchain. We have dedicated human resources in this area to solve problems of how to put contracts on the chain, and legal compliance for data supervision and monitoring. Because there is a lot of legislation and cyber security, and it involves cross-border data exchange, cryptography and other things".—Dr. Cao,

(continued)

Table 7.3 (continued)

Aggregate construct	2nd construct	1st construct	Solutions and supporting quotations
			BeefLedger researcher 2. To overcome the China firewall, it is necessary to transfer information and store it on a server in China. The legal department is also built for regulation issues. "Because there is a new legislation in China that requires that China's data must be stored in China, now we are also considering establishing one server in China, one in Australia".—Dr. Cao, BeefLedger researcher

et al. (2018): intra-organisational barriers, inter-organisational barriers, systems-related barriers, and external barriers.

7.3.1 *Intra-Organisational Barriers*

The intra-organisational barriers usually refer to the barriers related to an organisation itself. There are two main barriers underneath this category: define company demand and have strong determination, and financial constraints to implementing blockchain. All three companies either conducted research to gain knowledge about blockchain and the industry before launching the blockchain project, or ran multiple pilots to gain practical experience. All companies also invited experts from blockchain areas and the food industry area to discuss the decision of adopting blockchain. The financial constraints only applicable to the start-ups: BeefLedger and FairChain. Both received funding from governments to support their projects.

7.3.2 *Inter-Organisational Barriers*

The inter-organisational barriers refer to the collaboration issues between stakeholders. Blockchain is an information technology that requires more than one participant. Therefore, it is necessary to engage stakeholders, particularly suppliers. Companies first tried to invite the suppliers who are closer to them. With some pre-existing trust, it is easier to persuade suppliers. In addition, all companies also provided incentives to attract suppliers. W company committed to give better shelf place for the blockchain-traceable products. Moreover, companies also provided sufficient training to suppliers so they would understand the value of blockchain. Considering suppliers' financial constraints, W company covered all the costs. BeefLedger and FairChain also adopted cheaper data collection methods to reduce suppliers' financial pressure. The blockchain platforms are also designed to be supplier-friendly to fit the current supply chain system. The technology compatibility can save suppliers from replacing the whole system.

In addition, suppliers have concerns about business secret disclosure. To solve this issue, BeefLedger allows suppliers to choose what kind of information they want to share. W company only requires supplier performance data to be on-chain. For FairChain, the farmers are anonymous on-chain to protect farmers' data. BeefLedger and FairChain also had concerns about culture differences between stakeholders and focal companies. BeefLedger, therefore, conducted research to understand Chinese consumers' value. FairChain invited local NGOs and the community to educate suppliers.

The consumers are also given education via publicity by W company to increase consumer awareness to blockchain. Moreover, the blockchain platform is designed to be consumer-friendly by W company. Consumers can scan the QR code to connect with blockchain and check information. The convenience can largely encourage consumers to use blockchain. FairChain encourages consumers to expand the positive externalities and tip farmers by giving out tokens. Consumers can increase their awareness by participating in the blockchain project.

7.3.3 System-Related Barriers

Systems-related barriers include technology immaturity and environmental restrictions. As blockchain is a young technology, there are a few restrictions that need to be solved, such as raw data manipulation, scalability, and limited functions. To solve raw data manipulation, IoT devices are adopted by all case companies for automatic data collection. Multi-signature and triple checks before data uploading by BeefLedger and FairChain, respectively, can also enhance data authenticity. For FairChain, scalability is not a concern. However, BeefLedger is considering changing to a new infrastructure to solve potential scalability issues. W company combined on-chain and off-chain methods to make more space for important data.

Blockchain is only a fundamental infrastructure, so all companies integrated blockchain with IoT devices for better effectiveness.

The environmental restrictions are not applicable in the W company case. However, in the cases of BeefLedger and FairChain, there are many local suppliers in the areas with low digitisation. To enable blockchain, BeefLedger has helped their suppliers to build infrastructures. FairChain chose to collect data offline when the network is unstable, and upload data on-chain when the network is in good condition.

7.3.4 External Barriers

External barriers refer to the limitations caused by different policies. This is one of the side effects due to technology immaturity. Blockchain is a new technology; therefore, many countries have different opinions about it. In this case, BeefLedger found the China firewall restricted data-sharing between Australia and China. Therefore, Chinese consumers cannot reach data directly from the Australia side. To solve this issue, BeefLedger built a legal department targeting cross-border data-transferring. Moreover, BeefLedger also created a server in China to store data, which allows the Chinese consumers access to these data.

7.4 Project Outcomes

This section presents the case companies' blockchain project outcomes. Figure 7.4 and Table 7.4 provide the comparison of the three cases with supporting quotations.

7.4.1 Supply Chain and Traceability Efficiency

The supply chain efficiency can be largely improved by using blockchain. For BeefLedger, the permanently recorded product information can speed up the food recall process by isolating certain products. From the human resource utilisation perspective, as more supplier performance data are recorded on the chain, W company can manage their suppliers more efficiently from suppliers with higher risks to lower risks. From the financial perspective, FairChain redistributed their marketing budget and used the money to encourage consumers by sending out tokens.

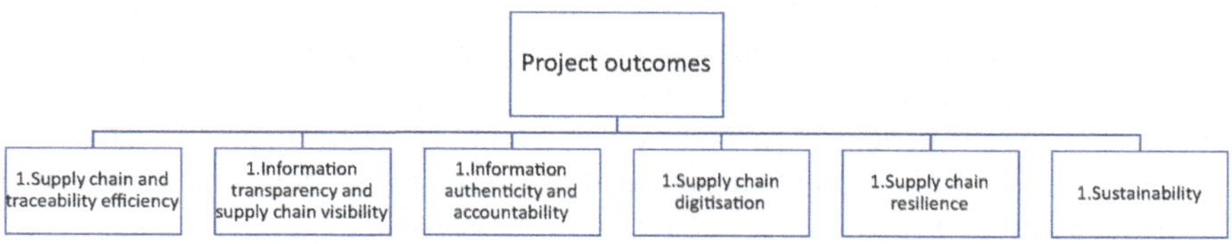

Fig. 7.4 Project outcomes

Table 7.4 Project outcomes of the case companies

Supply chain	BeefLedger	W Company	FairChain
Supply chain and traceability efficiency	Improving traceability efficiency: "You use sensors to obtain data, after you obtain it, you need someone to verify it. Retailers want to trace the source of responsibility. For example, to trace who is responsible for the product in each link in the process, if something goes wrong, it can be directly isolated".—Dr. Cao, BeefLedger researcher	Better supplier management: "If they have ten people to do factory inspection, for example, now they can rearrange tasks and re-allocate resources more effectively. For instance, they can reduce the inspection for lower risk suppliers from quarterly to semi-annually, and increase the site visiting frequency of higher risk factories. Therefore, the utilisation efficiency for personnel resources is greatly improved".—Mr. Cheng, the project manager in VeChain	Reallocating financial resource: "So instead of spending the marketing budget for brand into Facebook or Snapchat or you know, an advertisement running online or offline, but put that marketing money into hands of the consumers, so whenever they purchased a cup of coffee, they received a token".—Mr. A, the FairChain Architect.
Information transparency and supply chain visibility	Changing from linear to circulate supply chain: "Supply chains changed from chains to cycles; they are circuits, the economic circuits. So, every stage in the circuit has its own need for information and players who are engaged in each of those stages will perform different roles. The farmer can be validated in the network for information that is collected at a different point in the process. Nothing stopped somebody participating as a validator in the network. Nothing stops somebody running a storage node to store a copy of the information for the whole network".—Warwick Powell, Chairman of BeefLedger Streamlining information flow: "From the process perspective,	Enhancing supply chain transparency: "Early on, W company had to set up a complete supplier management standard, include access mechanism, regular audit and test. But there are unreliable factors such as cutting corners or quality issues. It was difficult for W company to obtain supplier performance data. Because W company only gets to know if a supplier is good or not when there is an incident. The cost to maintain a dedicated team to conduct regular inspections and spot checks on the suppliers is also huge".—Mr. Zhang, Co-founder of VeChain.	Bringing supply chain transparency: "That's why we call it radical transparency. The consumers will know in the value chain about who owns what. We've got the supply chain solution, which basically provides provenance of the product. And it also makes you follow the money. And you can look back from your product that you bought on who did earn what".—Mr. R, the chief technology officer of FairChain Foundation.

(continued)

Table 7.4 (continued)

Supply chain	BeefLedger	W Company	FairChain
	it can form a kind of interaction. For example, as a producer, he can understand the entire process of his product. In addition, he can understand the feedback from consumers, and to form an interactive mode. In many cases, Australia's traditional supply chain mode is fragmented in this way, and farmers receive insufficient and processed information. The information inconsistent and lagging with the market trend. Therefore, it is important to achieve real time Information flow, it will also help evidence based decision making".—Dr. Cao, BeefLedger researcher		
Information authenticity and accountability	Ensuring food authenticity: "From the perspective of the Australian industry, because Chinese and Australian beef is equivalent to the best quality in China, the possibility of fraud is also a lot higher. Blockchain is to strengthen a brand protection".—Dr. Cao, BeefLedger researcher	Ensuring information authenticity: "The product origins used to be manually entered, and can be changed in the background anytime. This kind of data changing is happening quite often in the food industry. For example, we all know that pears from Xinjiang are delicious, the suppliers can change the pear origins to Xinjiang. This is the weakness of traditional traceability method".—Mr. Feng, senior manager at PwC food & beverage retail.	Proving shared company efforts on shared value chain: "So what we want to see running backed up by blockchain is that the value distribution is clear for our consumers, because for us, it's important that we can prove that 50% of the value of the coffee as well. We sell a product, a bag of coffee for 1 kg, for 20 euros, something like that. And we want to demonstrate to our consumers that that 50% of that remains in Ethiopia. And that's why I use blockchain to have this immutable proof".—Mr. R, the chief technology officer

(continued)

Table 7.4 (continued)

Supply chain	BeefLedger	W Company	FairChain
			of FairChain Foundation.
Supply chain digitisation	Improving data checking efficiency: "For example, if you submit the documents at the customs, it will ask you to confirm, as your information has been uploaded on chain, and verified by multiple users, the customs will be able to recognize you, and let you pass quickly".—Dr. Cao, BeefLedger researcher	Enabling smart contract for a quicker payment: "For example, in the past, it usually has to conduct a multi-party verification to confirm the receiving of goods, and then payment can be initiated. The whole process is long. But if they start to use smart contract, as soon as suppliers ship the products and W company's distribution centres receive the products with the right amount and appropriate condition, the payment can be initiated. This is because the data of the products is recorded on the blockchain platform permanently. So, people has trust and no need to confirm so much things again. In other words, once the smart contract set up, we can synchronize the completion of the delivery and billing in both financial and legal sense".—Mr. Feng, senior manager at PwC food & beverage retail.	Enabling continuous auditing: "that's not in place yet, but that's one of the positive side effects of running this blockchain and that you're able to do a continuous audit thing instead of the every once a year an impact report auditing".—Mr. R, the chief technology officer of FairChain Foundation.
Supply chain resilience	Encouraging suppliers to take responsibilities: We have also implemented a multi-signatory process by which data is committed to the blockchain itself, which means that we ensure that participants have to take active responsibility for checking the validity of information that they submit. And that you need to have more than	Defining liabilities during incidents: "Before, suppliers give information to W company and let W company to display the information to consumers. By blockchain, as the data is immutable, both party can have more trust to each other. If there is a dispute, they can match the product data with the data on the chain to	

(continued)

Table 7.4 (continued)

Supply chain	BeefLedger	W Company	FairChain
	one person agree that the information is valid before it can go onto the blockchain".—Warwick Powell, Chairman of BeefLedger	define liability".—Mr. Cheng, the project manager in VeChain.	
Sustainability	Efficient food recall: although it is not the main focus on sustainability, but be able to track the cross-border beef supply chain, and conduct efficient recall when needed.	Digitising data to improve efficiency: "Instead of the traditional paper records, which is very inefficient with very little information on it, they (regulators) can see some detailed records on Blockchain if there is a quality accident".—Mr. Cheng, the project manager in VeChain	Promoting sustainability by improving supply chain transparency: "if our claim is that we contribute to living income and we improve the living income, you should be able to check that claim without asking for access to my data or something like that. And that's what we're currently working on with the team on how to make this public infrastructure available for all our consumers".—Mr. R, the chief technology officer of FairChain Foundation. Improving awareness of the consumers: "That's why we said shared value chain, it is not about us owning everything and just buying from the farmer. Now, the ambition is that all stakeholders in the value chain, including the consumers will be owning our value chain of coffee. That's what we now working on".—Mr. R, the chief

7.4.2 *Information Transparency and Supply Chain Visibility*

Blockchain is a distributed ledger and allows peer-to-peer information sharing. Thus, all verified users on blockchain share the same information in real time. This means that the information flow can be streamlined within a supply chain. For BeefLedger, the traditional linear supply chain can be changed to a circulate supply chain, which allows stakeholders to share and access product information at the same time. Not only the product information; W company can obtain supplier performance data. FairChain can share the payment receipts with the public to prove their achievements.

7.4.3 Information Authenticity and Accountability

The immutability feature of blockchain ensures that data become permanent once they are uploaded onto the chain. The blockchain data, therefore, can be trusted on their authenticity and accountability. This not only gives consumers buying confidence in the case of BeefLedger and W company, but also can be used as evidence to prove FairChain's efforts in promoting a sustainable coffee chain.

7.4.4 Supply Chain Digitisation

Blockchain can bring supply chain digitisation in several perspectives. For BeefLedger, beef product certifications can be digitised and recorded on blockchain to speed up the custom-checking process. In the case of W company, payment process can be executed automatically by enabling smart contract. For FairChain, the immutable data records can enable continuous auditing instead of annual reports.

7.4.5 Supply Chain Resilience

The company's supply chain resilience can be enhanced by blockchain. The supply chain transparency that blockchain brought can encourage stakeholders to take responsibilities for the data. Moreover, the immutable records can also define liabilities when food incidents happen.

7.4.6 Sustainability

Blockchain can also promote sustainability in a few ways. First, food wastes can be largely reduced due to improved traceability efficiency. Second, paper records can be replaced by recording data on the blockchain. Third, FairChain not only used the blockchain to prove their efforts on promoting sustainability, but also encouraged consumers to participate in the blockchain project to increase their awareness of sustainability. Consumers' awareness and participation can expand the positive externalities and have a longer impact on sustainability.

Overall, there are six benefits of blockchain. The features of decentralisation, immutability, security, and smart contract can largely enhance a supply chain's transparency and efficiency, reduce potential risks, and promote sustainability.

7.5 Summary

This chapter conducted cross-case analysis and compared the three cases in their decision-making process and blockchain implementation. There are four stages in between generating ideas and actual implementation (setting the stage activities, customer clue-gathering activities, negotiating, clarifying, and reflecting activities, and inter-organisational learning).

In addition, four key success factors (capabilities, collaboration, technology readiness, external environment) and eight sub-factors (knowledge, resources, leadership, stakeholder acceptance, technology maturity, technology compatibility, external support, market awareness) were suggested as the impact factors to blockchain implementation. Each of the factors is found to be crucial to the implementation process.

Moreover, following Saberi et al. (2018), common barriers in four categories (intra-organisational, inter-organisation, systems related, and external) were summarised. With the barriers, solutions taken by each company were also presented in this chapter.

To conclude, this chapter presented six main blockchain benefits to supply chain performance as the project outcomes (supply chain and traceability efficiency, information transparency and supply chain visibility, information authenticity and accountability.

References

Bermeo-Almeida, O., Cardenas-Rodriguez, M., Samaniego-Cobo, T., Ferruzola-Gomez, E., Cabezas-Cabezas, R., & Bazan-Vera, W. (2018). *Blockchain in agriculture: a systematic literature review*. 4th International Conference, CITI 2018, Guayaquil, Ecuador, November 6–9, 2018, Proceedings, 883, pp. 44–56.

Flint, D. J., Larsson, E., Gammelgaard, B., & Mentzer, J. T. (2005). Logistics innovation: A customer value-oriented social process. *Journal of Business Logistics, 26*(1), 113–147.

Kendall, H., Kuznesof, S., Dean, M., Chan, M., Clark, B., Home, R., Stolz, H., Zhong, Q., Liu, C., Brereton, P., & Frewer, L. (2019). Chinese consumer's attitudes, perceptions and behavioural responses towards food fraud. *Food Control, 95*, 339–351.

Saberi, S., Kouhizadeh, M., Sarkis, J., & Shen, L. (2018). Blockchain technology and its relationships to sustainable supply chain management. *International Journal of Production Research, 57*(7), 2117–2135.

Su, S. I. I., Gammelgaard, B., & Yang, S. L. (2011). Logistics innovation process revisited: Insights from a hospital case study. *International Journal of Physical Distribution & Logistics Management, 41*(6), 577–600.

Van Hoek, R. (2019). Exploring blockchain implementation in the supply chain. *International Journal of Operations & Production Management, 39*(6–8), 829–859.

Chapter 8
Discussion

This chapter discusses and critiques the research findings and prior related research. The innovation process model (Flint et al., 2005; Su et al., 2011) and the practice-based view (Bromiley & Rau, 2014) are applied, respectively. Several propositions are then formulated based on the discussions.

8.1 Innovation Process Model

The innovation process model suggested that there are four stages during a company's decision-making on innovation adoption (Flint et al., 2005). The model emphasises the process of why a company decides to adopt an innovation and how it proceeds. Thus, the four stages (setting the stage activities, customer clue-gathering activities, negotiating, clarifying and reflecting activities, and inter-organisational learning) present the process from idea generation to actual implementation. The 'setting the stage activities constitute the first stage, in which an interactive scene was set to allow initial ideas to emerge informally (Bjorklund & Forslund, 2018; Flint et al., 2005). The next stage is to collect the needed information to achieve a better view of the initial ideas (Flint et al., 2005). The third stage is an in-depth repetitive discussion stage with selected stakeholders, as well as implementation (Flint et al., 2005; Su et al., 2011). The final stage is to enlarge the scope of participants and optimise the innovation to be more user-friendly (Bjorklund & Forslund, 2018; Flint et al., 2005; Su et al., 2011).

C. Zhang et al., *Blockchain Applications in Food Supply Chain Management*,
Contributions to Management Science,
https://doi.org/10.1007/978-3-031-27054-3_8

8.1.1 Setting the Stage Activities

The first stage is *setting the stage* activities. According to Flint et al. (2005, p. 127), setting the stage activities “are those activities designed to create an environment conductive to interacting with and listening to customers and being an innovative organization”. At this stage, companies are creating an interactive environment for key business partners and have a few preparations to lay a solid foundation before taking further steps (Flint et al., 2005). Su et al. (2011) also suggested that the directions of the project can be clearer after interactions take place between business partners. Therefore, the first stage is for the case companies and their business partners to discuss the company issues and demand.

Song et al. (2020), van Hoek (2019) and Verhoeven et al. (2018) suggested the importance of diagnosing company issues and clarifying company demand before adopting innovations. Thus, at this stage, a few issues were identified by the case companies, such as fragmented information flow, supplier management efficiency, and value chain inequality. Following the issues, company demands were also proposed, targeting these issues. From the observations, different characteristics of different supply chains determine the different company needs that motivate them to adopt blockchain. The food supply chain is vulnerable and complex, and it is expected to be revolutionised by blockchain in the future (Kshetri, 2018; Rogerson & Parry, 2020). There are many forces driving the case companies to implement blockchain that can also be found in multiple research papers; these forces include to enhance consumer buying confidence, to improve supply chain transparency, to better define responsibilities and liabilities during food scandals, to be more efficient in supplier management, to digitise records, and to be used as a story-proving tool (Hua et al., 2018; Tieman & Darun, 2017; van Hilten et al., 2020; van Hoek, 2019; Yiannas, 2018). For example, BeefLedger and W Company both proved that consumers have higher requirements on food authenticity and traceability on certain food products such as luxury food products (Tian, 2016). This research finding also agrees with Rogerson and Parry (2020) and van Hoek (2019) who found that customer demand and product security are the key factors to motivate company blockchain adoption. According to Pearson et al. (2019), blockchain systems are most likely to be applied for some food products where most needed.

In addition to identifying issues and establishing demand, all the case companies formed strategic partnerships with experts from related industries. This observation is in line with Wagner (2008), who emphasised the importance of having industry experts in innovation to acquire sufficient knowledge as some literature claimed that the initial costs for blockchain implementation are the barriers to SMEs (Leong et al., 2018; Pearson et al., 2019; Perboli et al., 2018; van Hoek, 2019). Thus, for SMEs and start-ups, such as BeefLedger and FairChain, financial support from third parties is extremely important. Combining the observations and the literature, the first proposition is formulated:

Proposition 1 *Identifying issues and clarifying business demand based on the characteristics of the supply chain positively relates to blockchain adoption.*

8.1.2 Customer Clue-Gathering Activities

The second stage is *customer clue-gathering* activities, which include activities of interacting with customers, managing customer groups, engaging with customers through in-depth interviews, holding customer retreats, holding planning meetings, and conducting customer research (Flint et al., 2005). Su et al. (2011) further verified and suggested that this stage can also be interpreted from the perspectives of suppliers and stakeholders. Before decision-making and actual implementation, this stage is to help companies gain the necessary knowledge and experience to avoid failures (Chapman et al., 2002; Wagner, 2008). Therefore, companies are collecting data and testing various business plans to find the best fit. All three companies also conducted pilots to gain practical experience.

Many researchers suggested that lack of understanding is one of the challenges for blockchain implementation, and it is suggested that companies have a comprehensive understanding of the value of the technology before blockchain implementation (Leong et al., 2018; Verhoeven et al., 2018; Hackius and Petersen, 2017; Galvez et al., 2018; Queiroz & Wamba, 2019; van Hoek, 2019; Zhao et al., 2019). Autry and Griffs (2008), Flint et al. (2008), and Grawe (2009) found a positive relationship between knowledge and innovation. This observation is in line with the literature above; it is necessary for companies to carry out thorough research before making decisions. The research can be carried out in various ways, either via academic research or piloting.

On top of the data collection, it is suggested that a feasible business model is necessary before actual implementation. All three cases suggest that the real implementation may only take a few months; however, the design can impact long-term development. Moreover, Dutta et al. (2020) claimed that the actual implementation is easy as it usually requires multiple parties to get involved, and many challenges need to be overcome. van Hoek (2019) also expressed the same concern as piloting is quicker because of the small scope. Thus, the second proposition is formulated:

Proposition 2 *Sufficient knowledge of the value of blockchain positively relates to blockchain adoption decision-making and implementation.*

8.1.3 Negotiating, Clarifying, and Reflecting Activities

The third stage is *negotiating, clarifying, and reflecting* activities. This stage is described by Flint et al. (2005) as a phase of conducting internal negotiation, clarifying the demand of consumers, and deciding the key aspects of the innovations. In the case study by Su et al. (2011), the third stage was to design innovation based on prior research and knowledge and collect the opinions of both the focal company and the suppliers. From observations, this stage is mainly about designing and implementing blockchain based on the experience from pilot studies and knowledge from researchers and roll it out to engage key stakeholders.

Although many researchers suggested that blockchain can help to form trust between stakeholders (Kshetri, 2018; Rogerson & Parry, 2020), the finding suggests that companies tend to have a certain level of trust before implementing blockchain. This is in line with van Hoek (2019), who suggested that blockchain implementation could start quickly with some selected business partners. From the observation of the three case companies, all companies have a few years of cooperation with their suppliers during blockchain implementation. In this case, a certain level of trust between case companies and suppliers was already in place. This observation is in line with Van Hilten et al. (2020), who suggested that companies tend to adopt blockchain within a less complex supply chain or within a well-arranged supply chain with certain trust. It is believed that a close relationship between partners can make blockchain implementation faster and more efficient (Van Hilten et al., 2020).

Many researchers found the importance of networks in innovation (Chapman et al., 2002; Grawe, 2009; Hakansson & Persson, 2004). Collaboration between supply chain partners is considered a key influencing factor in blockchain implementation (Dutta et al., 2020; Hastig & Sodhi, 2020; Post et al., 2018; Tan et al., 2018). The observation of the case companies also suggests the importance of supplier engagement. All three companies made efforts to persuade the suppliers to accept blockchain by providing training and reducing suppliers' financial pressure. The research on supply chain technology by Lin et al., (2016) proposed that firms are motivated by the perceived benefits and tend to invest in training programmes on supply chain technologies. Wagner (2008) also agreed that training is one of the most important innovation activities. Education on blockchain knowledge can help companies to persuade suppliers to join the project (Tan et al., 2018).

Moreover, Tan et al. (2018) proposed that wide adoption of blockchain can only happen when participants are strongly motivated by value propositions. It is also important to make the blockchain system easy to use with affordable costs for smaller businesses (Leong et al., 2018; Pearson et al., 2019; Perboli et al., 2018). Song et al. (2020) suggested that stakeholders' demands and expectations are essential during blockchain building. The leadership commitment was found to be positively related to blockchain implementation (Van Hoek, 2019). Kshetri (2018) also found that the influence that a focal company can bring to suppliers' blockchain adoption is important. The observation found that suppliers' financial pressure is one of the influencing factors for supplier acceptance of the new technology. Thus, the case companies adopted different ways to reduce the costs for suppliers and kept the blockchain system easier to adopt. Thus, the third proposition is formulated.

Proposition 3a *The pre-existing trust among supply chain partners is positively associated with the smooth blockchain adoption.*

Proposition 3b *Supplier training and education are positively associated with the blockchain implementation process.*

8.1.4 Inter-Organisational Learning

Inter-organisational learning is the last stage of the innovation process. This is a stage of new understanding and insights, as well as potentials, possibilities, and future directions (Flint et al., 2005; Su et al., 2011). The new insights emerge from the joint learning of both the focal company and stakeholders. From the observations, this is the phase to collaborate with more stakeholders and remain flexible on systems changing.

Rogerson and Parry (2020) found that the potential value of blockchain is reduced if the blockchain application is not available to most people. Sander et al. (2018) also suggested that if the blockchain platform is "consumer-friendly" this could encourage consumers to use the platform. Thus, to invite more stakeholders on board, all the case companies tried to optimise the blockchain system to make it as user-friendly as possible. Leong et al. (2018) and Pearson et al. (2019) also suggested that blockchain should be "SME friendly" on technology accessibility. Partially replacing the existing system rather than the whole existing system by blockchain can save significantly more costs for small businesses (Perboli et al., 2018; van Hoek, 2019). The case companies, therefore, designed the blockchain platform to be combined with suppliers' existing system without many extra facilities.

Leong et al. (2018) suggested that different stages of the supply chain might have different requirements on technology adoption. The company demand may change over time and companies should remain flexible to changes. The observation also found that the case companies kept updating the blockchain platform and "learning by doing". Various new functions were added to the blockchain platform to make it more efficient and user-friendly. Thus, the fourth proposition is formulated:

Proposition 4 *Constant blockchain system adjustment in line with the changes of demand and situation positively relates to blockchain implementation.*

8.2 Blockchain Adoption Influencing Factors and Performance Outcome

In contrast with the resource-based view (RBV), the practice-based view (PBV) emphasises the imitable and transferable practices that influence company performance. "The practice-based approach stresses how resources (e.g., tools, images and language) are implemented and integrated as a part of everyday life through actions and interactions" (Russo-Spena & Mele, 2012, p. 538). Following the innovation process model, the PBV presents the supply chain performance improvement after blockchain implementation. The PBV provides an observation lens to the details of the innovation process and helps to summarise the commonalities between the firms during the innovation process. Moreover, Carter et al. (2017) proposed a supply chain practice view (SCPV) which expands the PBV from an intra-organisational level to an inter-organisational level of analysis. SCPV involves the practices that happen in more than one firm. Thus, in the supply chain scenario, the practices of both case companies and suppliers are analysed. This section combines the PBV and SCPV to summarise the key success factors, the common barriers faced by the three case companies during blockchain implementation, and the performances achieved during blockchain adoption.

8.2.1 Critical Success Factors

Hastig and Sodhi (2020) proposed six critical influencing factors for blockchain implementation: capabilities, collaboration, technological readiness, supply chain practices, leadership, and governance of traceability effort. Yadav and Singh (2020) identified six major reasons for a company adopting blockchain: data safety and decentralisation, accessibility, laws and policy, documentation, data management, and quality. Post et al. (2018) suggested 13 factors that influence blockchain diffusion based on expert interviews: necessary collaboration, necessary paradigm shift, market position adoption, compliance, sector pressure, organisational size, investment hesitation, knowledge deficit, viable use cases, implementation method, change readiness, technical shortcomings, and process maturity. Saurabh and Dey (2021) suggested six factors that contribute to agri-food company blockchain adoption: traceability, dis-intermediation, trust, coordination/control, compliance, and price of technology products/services. This research identifies four categories and eight critical success factors that combine both prior research findings and observations from the case companies.

8.2.1.1 Capabilities

Porter (1996) suggested that even though the firms engage in similar practices, the results may still vary due to different capabilities. Thus, capability is one of the influencing factors to blockchain implementation. Capabilities generally mean a company's abilities that include knowledge and resources to implement the blockchain project.

Knowledge

Kouhizadeh et al. (2021) suggested that organisational knowledge is one of the key factors in capabilities. From the observation, before taking further steps, all companies have gained a certain level of understanding of the industry and blockchain, either by conducting research, running pilots, or attending industry conferences. Companies can learn a meaningful lesson from piloting before officially launching blockchain (van Hoek, 2019) For example, BeefLedger carried out a year-long research study before launching the project. Similarly, W company had pilot experience as early as 2016. FairChain also teamed up with a blockchain expert to trace coffee from 2018.

Resources

A company's resources can also influence blockchain implementation (Kouhizadeh et al., 2021). The resources include all kinds of tangible and intangible resources such as financial resources and human resources (Kouhizadeh et al., 2021). Apart from W company, which is a world-leading retailer that has a strong financial background, BeefLedger and FairChain have received government funding to run their projects. From a human resource perspective, all three companies cooperated with industry and technology experts to jointly complete the projects.

8.2.1.2 Collaboration

Collaboration refers to the trust and acceptance of supply chain stakeholders for the blockchain project. It was suggested as one of the key factors in innovation by many researchers (Chapman et al., 2002; Grawe, 2009; Hakansson & Persson, 2004). From the observations, collaboration can largely influence the progress of the blockchain project.

Leadership

Leadership is a company's ability to bring stakeholders on board with the blockchain project. A focal company's decision on blockchain adopting can influence their stakeholders' behaviours relating to blockchain adoption (Kouhizadeh et al., 2021; Kshetri, 2018; van Hoek, 2019). For example, in the case of BeefLedger and FairChain, some suppliers suffer from low digitalisation and poor network connectivity. BeefLedger led by example to motivate suppliers and help to improve local digitalisation where necessary. W company and FairChain covered all the investments for suppliers. W company selected a cheap data collection method—the QR code— to reduce suppliers' financial pressure. Many of FairChain's suppliers are living under the poverty line and do not have a smartphone. FairChain provides farmers with various forms of training to help them earn more income.

Stakeholder Acceptance

From the very beginning, all companies chose to invite suppliers who have close relationships or share similar goals with them. The pre-existing trust can speed up the stakeholder acceptance progress. In addition, incentives were also provided to attract suppliers. BeefLedger employed professionals to communicate with suppliers to check their attitude to blockchain. Sufficient training was also provided by all companies to their suppliers. Moreover, the three companies adopted data collection methods that are cheap and accessible for suppliers, such as RFID and QR code. This can also encourage suppliers to accept the blockchain platform.

8.2.1.3 Technology Readiness

The technology readiness includes the maturity of blockchain and its compatibility with the current company system.

Technology Maturity

Blockchain is a young technology with some technical restraints. Kouhizadeh et al. (2021) suggested that blockchain immutability can cause concerns for managers to adopt blockchain. For instance, scalability is one of the technological issues. W company chose to keep most information off-chain and leave the responding hash on chain. FairChain expressed a positive attitude to the sociability issue. Another issue relates to data mistakes during data uploading. BeefLedger enabled a multi-signature function to verify data before allowing data to be uploaded. W company not only invited third parties to check information but can also temporarily hide wrong information from consumers to avoid potential conflicts. In the case of FairChain, the data should be checked three times before data uploading. In addition, all three companies adopted IoT devices to collect data. IoT devices are perfectly compatible with the blockchain platform and help to collect accurate data (van Hoek, 2019).

Technology Compatibility

Technology compatibility refers to how blockchain can fit company demand without interrupting current operating systems. van Hoek (2019) mentioned that interoperability between different systems may be a barrier. Thus, in the case of BeefLedger and W company, the blockchain platforms were designed to be as user-friendly as possible for suppliers to adopt. Suppliers were not required to invest in the new system, or to replace their facilities. The blockchain platform can easily fit in their daily operating routine. In the FairChain case, most farmers do not own a smartphone. Thus, the largest investment for farmers is purchasing smartphones.

8.2.1.4 External Environment

External environment refers to the external factors that can influence blockchain implementation progress. The factors include the support from third parties and the market awareness of blockchain.

External Support

The government attitude to blockchain adoption can strongly influence a company's decision-making. Kouhizadeh et al. (2021) suggested that pressure from the external environment can be the motivation for companies to adopt blockchain in order to fall in line with external expectations. Sander et al. (2018) also said that the government should step in to encourage stakeholders to use blockchain to share information if necessary. All three companies received different levels of support from the government at the beginning of the project. For example, BeefLedger and FairChain applied for funding from the government to support their projects while W company built a close relationship with local government.

Market Awareness

The market reaction to blockchain is also an influencing factor. Nowadays, consumers tend to know more about product information when making purchases (Casado-Vara et al., 2018; Rogerson & Parry, 2020). From the observation, the Chinese government emphasises the importance of adopting new technology to ensure food safety. These changes from the external environment also potentially influence the market attitude to blockchain. Thus, the fifth proposition is formulated:

Proposition 5 *A company's capabilities, collaboration level, technology readiness, and external environment positively relate to blockchain implementation.*

8.2.2 Common Barriers and the Solutions

Imitable practices are the practices that can be transferred between firms without strong isolating mechanisms (Bromiley & Rau, 2014). Therefore, this section presents the common barriers encountered during blockchain implementation and the solutions employed by the case companies. The section adopts the four categories of barriers by Saberi et al. (2018): intra-organisational barriers, inter-organisational barriers, systems-related barriers, and external barriers.

8.2.2.1 Intra-Organisational Barriers

The intra-organisational barriers concern the barriers within a company and mostly happen in the first two stages of the innovation process. The companies in the setting-up stage are identifying issues and demand and gathering business partners. Thus, the first barrier at this stage is to identify company demand before decision-making. All three case companies had invited strategic partners and held various meetings to discuss the company demand. Multiple research studies were conducted along with running pilot studies before officially kicking off the project. The practical experience from running pilots can deepen knowledge and understanding of blockchain technology.

The financial constraint to investing in an innovation is another barrier mentioned by a few researchers, particularly for SMEs (Ali et al., 2021; Leong et al., 2018; Pearson et al., 2019; Perboli et al., 2018; Tan et al., 2018; van Hoek, 2019). As start-ups, therefore, BeefLedger and FairChain had applied for government funding to support their projects. On the other hand, as one of the world's largest retailers, W company has no such concerns. W company not only funded the blockchain project, but also covered all the costs including consulting costs and supplier training costs in order to speed up the project.

8.2.2.2 Inter-organisational Barriers

The inter-organisational barriers are the barriers between stakeholders. In this research, the barriers mostly happen in the last two stages of the innovation process. The negotiating, clarifying, and reflecting stage is when the direct participants of the project—the suppliers—are invited. Therefore, the barriers are all about the interactions with suppliers, including how to attract suppliers to participate in the project, how to educate suppliers, and how to solve financial pressures for suppliers. Many researchers suggested that collaboration is a challenge for blockchain implementation (Dutta et al., 2020; Kshetri, 2018). To attract suppliers, all the companies provided different incentives to suppliers. For example, W company promised the suppliers who participated in the blockchain project that they can list their products in better shelf places. BeefLedger hired professionals to check the suppliers' acceptance of the new technology and provided training education to the suppliers. FairChain tried to build a close relationship with the suppliers by raising the purchasing price and financing a wet mill. It is also noticeable that all companies first invited the suppliers who have the closest relationships with them. In other words, the existence of a certain level of trust can help companies to launch innovation.

With the target of providing suppliers with sufficient knowledge and understanding of blockchain technology, all the companies arranged either meetings with or on-site guidance to their suppliers. Sufficient knowledge of blockchain can also help companies to persuade suppliers to join in the project. Many suppliers have concerns over the fees involved to install the blockchain application or replace the infrastructure. To solve this concern, all companies designed the blockchain system to be affordable and easy to use. IoT devices were carefully decided on based on the functionality and costs. For example, W company chose the QR code over FRID due to costs.

In the stage of inter-organisational learning, the barriers are the cultural difference between stakeholders and company, and lack of awareness of consumers. The cultural differences are the barriers mainly for BeefLedger and FairChain, as W company sources most of its products locally. BeefLedger targets the Chinese market while producing products in Australia. Therefore, BeefLedger had conducted research and held roadshows in China to understand Chinese consumers. Similarly, FairChain sources coffee in Ethiopia but the company is based in the Netherlands. FairChain solved the language barriers and cultural difference by inviting NGOs and local teams to communicate with farmers.

Another inter-organisational barrier is the lack of consumer awareness and engagement. This barrier was also mentioned by Rogerson and Parry (2020). Targeting this barrier, BeefLedger and W company designed their interfaces to be more user-friendly, which was intended to lower the entry level for consumers. FairChain encouraged consumers' participation by giving out tokens to jointly create a shared value chain.

Stakeholders may have concerns on data privacy during data sharing, which is also a barrier (Kouhizadeh et al., 2021; Sander et al., 2018). For all three companies, suppliers are not required to upload their sensitive information on to the chain. Moreover, BeefLedger gives suppliers the freedom to choose the information they prefer to upload. Similarly, W company only requires supplier performance data and products' quality data. FairChain anonymised the farmer data and stored sensitive data off-chain.

The last systems-related barrier relates to limited digitalisation in some areas in the case of BeefLedger and FairChain. To solve this barrier, BeefLedger developed a data interface that works well for basic CSV and PDF files. FairChain combined the online and offline data collection method. W company has not encountered this issue yet as most suppliers have good network connectivity. Some of the least developed areas may not be ready to adopt blockchain; this is also mentioned in the research by Kshetri (2018) and Dutta et al. (2020).

8.2.2.3 Systems-Related Barriers

The systems-related barriers relate to the technology and mostly happen at the third stage of the innovation process. The third stage is the negotiating, clarifying, and reflecting stage, which also includes blockchain system design and implementation. At this stage, all companies faced a few technological barriers. The first barrier relates to the limited function of the blockchain. Blockchain is only a fundamental infrastructure, which usually works with other devices for better efficiency. Kshetri (2018) also expressed concerns over the possibilities of raw data manipulation. Thus, all three companies combined their blockchain systems with IoT devices for accurate data collection. van Hoek (2019) also suggested that IoT devices and blockchain are complementary.

The scalability issue was mentioned by a few researchers (Pearson et al., 2019). It is also another barrier for BeefLedger and W company. W company combined on-chain and off-chain methods to reduce the size of the data uploaded on-chain. BeefLedger also requires a large storage space and will focus on solving the scalability issue in the future. However, FairChain does not see scalability as a problem and believes it can be solved along the blockchain development. For the barrier concerning data mistakes, all the companies deployed IoT devices to increase data accuracy during data collection. In addition, BeefLedger enabled a multi-signature scheme to verify data. FairChain also invited its partners to check their data before data uploading. W company can hide the wrong information from consumers in an emergency if there are data mistakes.

8.2.2.4 External Barriers

The external barriers are the barriers about external organisations or external stakeholders. These barriers are mainly present in the last stage of the inter-organisational learning stage. In this stage, companies are including more stakeholders in the projects but are also restricted by some policies. Kshetri (2018) proposed that the challenges relate to different laws and regulations. From the observation, BeefLedger is facing China's firewall, which does not allow Chinese consumers to connect directly to the blockchain system in Australia. Therefore, BeefLedger not only created a server on the China side that stores information, but also built a legal department to target the issues in the legal area. W company is restricted to collecting consumer information. They are also seeking advice from legal departments to align with policy requirements. Thus, proposition 6 is formulated:

Proposition 6 *There are usually four types of barriers that need to be overcome during blockchain implementation.*

8.2.3 Project Outcomes

The PBV emphasises performance improvements after adopting certain practices. The positive relationship between innovation and firm performance was suggested by many researchers (Chapman et al., 2002; Grawe, 2009). Among the innovations, blockchain was found to be able to bring various benefits to the supply chain (Alkhudary et al., 2020; Dutta et al., 2020; Faye, 2017; Feng et al., 2020; Galvez et al., 2018; Howson, 2020; Lin et al., 2017; Lin et al., 2020; Song et al., 2020; Tian, 2016; van Hilten et al., 2020). The observation of the case companies also agrees with the extant studies about blockchain benefits. This section presents the blockchain benefits that are found from case companies, and a critique of extant studies. There are six key benefits: supply chain and traceability efficiency; information transparency and supply chain visibility; information authenticity and accountability; information authenticity and accountability; supply chain digitisation, and sustainability. Different case companies accrued different blockchain benefits based on the company demands and supply chain characteristics.

8.2.3.1 Supply Chain and Traceability Efficiency

The human resource utilisation efficiency can be significantly improved by the blockchain system in the case of W company. From the observation, the blockchain system allows W company to gain more supplier performance data and encourage suppliers to behave honestly. Thus, compared to the traditional way of supplier auditing and factory inspection, W company can reallocate labour resources to the areas where they are needed. For instance, W company can pay more attention to factory auditing and inspecting to prevent potential risks for certain suppliers. This is in line with Tieman and Darun (2017) and Tan et al. (2018), who claimed that companies can choose suppliers and business partners based on the trading history that is stored on the chain.

From the financial perspective, instead of investing in marketing, FairChain enabled tokenisation to engage consumers in promoting sustainability. Consumers are encouraged to contribute to the shared value chain by tipping farmers with tokens. The sense of participation not only enhances consumer loyalty but also attracts new consumers. Rather than believing in what a company says, consumers can join in the project to help farmers. Thus, FairChain can save money from the marketing budget and spend it on some more valuable things.

Moreover, the literature suggested that blockchain is helpful in traceability efficiency (Lin et al., 2018; Yiannas, 2018). From the observations, W company also suggested the usefulness of blockchain in traceability. The immutable data records and real-time information monitoring can largely reduce the amount of time needed for food tracing and recall.

8.2.3.2 Information Transparency and Supply Chain Visibility

For BeefLedger, blockchain has optimised the information flow by turning the linear supply chain into a circular supply chain. Stakeholders are sharing real-time information and avoiding information asymmetry. In the case of W company and FairChain, supply chain transparency can also be achieved by blockchain. W company can obtain supplier performance data as well as product information at the same time. FairChain can prove their performance on achieving sustainability by sharing with the public the costs and payments involved in each stage of the coffee value chain.

The above observation agrees with the finding by Lin et al. (2016), who suggested that supply chain technology utilisation positively relates to supply chain information sharing and delivery performance. Galvez et al. (2018) and Kshetri (2018) claimed that supply chain transparency can influence consumer buying confidence and supply chain performance. Rogerson and Parry (2020) also found that blockchain can offer supply chain visibility and reduce supply chain friction.

8.2.3.3 Information Authenticity and Accountability

Moreover, each product is assigned a unique token to prove product authenticity. Consumers can have confidence while buying the products with the token. For both BeefLedger and W company, blockchain brings information transparency to satisfy consumers' demand for information. Consumers are more confident during food shopping. For FairChain, the immutability feature of blockchain can be used to prove its achievements in shared value chain and sustainability. Immutable records can enhance consumer trust in the company and the products. The findings are consistent with existing research that the immutability feature of blockchain can bring benefits to the supply chain (Rogerson & Parry, 2020). Blockchain-based transactions are auditable (Kshetri, 2018). Companies can use innovation as a marketing tool to differentiate them in the market and attract purchases (Grunert, 2005; Mai et al., 2010). Blockchain can be used to improve company image, attract new consumers, and enhance existing consumer loyalty (Galvez et al., 2018). Sander et al. (2018) and Stranieri et al. (2021) suggested a positive influence of blockchain on consumer-educated purchasing decision-making and satisfaction.

8.2.3.4 Supply Chain Digitisation

By combining smart devices such as IoT, blockchain allows data to be collected automatically and recorded permanently on-chain instead of written on papers. Digitised recording not only saves space from paper records but also improves data efficiency and security. Keeping data on blockchain reduces the possibility of data loss and data alteration and allows stakeholders to keep an eye on information flow. The monitoring of information flow is particularly important for frozen and fresh products in order to maintain product freshness and quality (Lin et al., 2018; Tian, 2016, 2017). The data can be used for quicker and easier custom document-checking for BeefLedger and continuous auditing for FairChain. In addition, applying a smart contract in the supply chain can help to achieve delivery and billing synchronisation by enabling fast and secure payment. The smart contract can speed up the payment process by automatically executing itself when certain conditions are met.

8.2.3.5 Supply Chain Resilience

Another performance improvement is risk reduction. Risks are the factors that negatively relate to supply chain performance (Wamba et al., 2020). Blockchain application can help companies to reduce risks by the immutability feature (Galvez et al., 2018). The immutable records can help companies to isolate problematic products quickly during food recall to reduce negative effects. In addition, the immutable record encourages suppliers to behave honestly which can prevent dishonest behaviours from the beginning. It also defines responsibilities during food scandals and eliminates possible disputes between companies. In other words, blockchain can be a tool for companies to better manage their supply chains and to improve their supply chain governance (Stranieri et al., 2021).

8.2.3.6 Sustainability

Blockchain is found to be helpful in improving sustainability in different ways (Rogerson & Parry, 2020; Song et al., 2020; Tsolakis et al., 2020; van Hoek, 2019; Yiannas, 2018). Real-time information monitoring can also help supply chain stakeholders to react to certain situations quicker without influencing the whole supply chain, such as food recall (van Hoek, 2019). In the case of W company, by keeping immutable information, blockchain can help to accelerate the food recall process and target the problematic food products as soon as possible to reduce negative effects and food wastes during food recall.

Apart from the environmental perspective, blockchain also helps to promote sustainability from economic and social perspectives by awakening awareness of the consumers. For FairChain, the tokenisation function can be used to encourage consumers to participate in the equal value chain and extend the positive externalities. Moreover, the costs and payment information of each coffee supply chain stage are also recorded on blockchain to support FairChain's efforts in promoting equality. Thus, proposition 7 is formulated:

Proposition 7 *Blockchain implementation positively relates to supply chain performance by improving supply chain efficiency, transparency, accountability, digitisation, resilience, and sustainability.*

8.3 Summary

This chapter presents a cross-case analysis, which is based on three within-case analyses and prior research. The chapter discussed blockchain applications in the food supply chain and answered the research questions through the innovation process model and the practice-based view. Thus, seven propositions (Table 8.1) and a conceptual framework (Fig. 8.1) are proposed. The observations and

Table 8.1 A summary of propositions

Number	Propositions
P1	Identifying issues and clarifying business demand based on the characteristics of the supply chain positively relates to blockchain adoption.
P2	Sufficient knowledge of the value of blockchain positively relates to blockchain adoption decision-making and implementation.
P3	a. the pre-existing trust among supply chain partners is positively associated with smooth blockchain adoption. b. Supplier training and education is positively associated with blockchain implementation process.
P4	Constant blockchain system adjustment in line with the changes of demand and situation positively relates to blockchain implementation.
P5	A company's capabilities, collaboration level, technology readiness, and external environment positively relate to blockchain implementation.
P6	There are usually four types of barriers that need to be overcome during blockchain implementation.
P7	Blockchain implementation positively relates to supply chain performance by improving supply chain efficiency, transparency, accountability, digitisation, resilience, and sustainability.

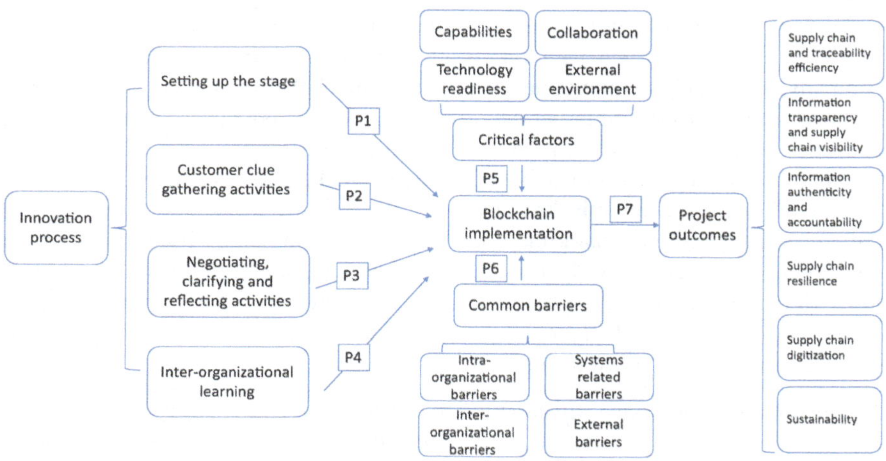

Fig. 8.1 Proposed framework of innovation process and practice-based view of blockchain implementation

discussions agree with the extant literature that companies should clearly identify the company demand (supported by P1) and gain sufficient knowledge (supported by P2) before making decisions to adopt blockchain. During blockchain implementation, the existing trust and stakeholder engagement can significantly impact the implementation process (supported by P3a and P3b). With the changes in needs, companies should maintain flexibility in adjusting the blockchain system (supported by P4). There are four key success factors (supported by P5) and four key barriers are found from observations (supported by P6). The observations also confirm that blockchain application can improve supply chain performance (supported by P7).

References

Ali, M. H., Chung, L., Kumar, A., Zailani, S., & Tan, K. H. (2021). A sustainable blockchain framework for the halal food supply chain lessons from Malaysia. *Technological Forecasting and Social Change, 170*, 1–14.

Alkhudary, R., Brusset, X., & Fenies, P. (2020). Blockchain in general management and economics: A systematic literature review. *European Business Review, 32*(4), 765–783.

Autry, C. W., & Griffs, S. E. (2008). Supply chain capital: The impact of structural and relational linkages on firm execution and innovation. *Journal of Business Logistics, 29*(1), 157–174.

Bjorklund, M., & Forslund, H. (2018). Exploring the sustainable logistics innovation process. *Industrial Management & Data Systems, 118*(1), 204–217.

Bromiley, P., & Rau, D. (2014). Research prospective: Towards a practice-based view of strategy. *Strategic Management Journal., 35*(8), 1249–1256.

Carter, C. R., Kosmol, T., & Kaufmann, L. (2017). Toward a supply chain practice view. *Journal of Supply Chain Management, 53*(1), 114–122.

Casado-Vara, R. C., Prieto, J., Prieta, F. D. L., & Corchado, J. M. (2018). How blockchain improves the supply chain: Case study alimentary supply chain. *Procedia Computer Science, 134*, 393–398.

Chapman, R. L., Soosay, C., & Kandampully, J. (2002). Innovation in logistic services and the new business model: A conceptual framework. *Managing Service Quality, 12*(6), 358–371.

Dutta, P., Choi, T., Somani, S., & Butala, R. (2020). Blockchain technology in supply chain operations: Applications, challenges and research opportunities. *Transportation Research Part E, 142*, 1–33.

Faye, P. S. (2017). *Use of blockchain technology in agribusiness: Transparency and monitoring in agricultural trade*. In Advances in Economics, Business and Management Research, Proceeding of the 2017 International Conference on Management Science and Management Innovation (MSMI 2017), 31, pp. 38–40.

Feng, H., Wang, X., Duan, Y., Zhang, J., & Zhang, X. (2020). Applying blockchain technology to improve Agri-food traceability: A review of development methods, benefits, and challenges. *Journal of Cleaner Production, 260*, 1–15.

Flint, D. J., Larsson, E., & Gammelgaard, B. (2008). Exploring processes for customer value insights, supply chain learning and innovations: An international study. *Journal of Business Logistics, 29*(1), 257–281.

Flint, D. J., Larsson, E., Gammelgaard, B., & Mentzer, J. T. (2005). Logistics innovation: A customer value-oriented social process. *Journal of Business Logistics, 26*(1), 113–147.

Galvez, J. F., Meijuto, J. C., & Simal-Gandara, J. (2018). Future challenges on the use of blockchain for food traceability analysis. *Trends in Analytical Chemistry., 107*, 222–232.

Grawe, S. J. (2009). Logistics innovation: A literature-based conceptual framework. *The International Journal of Logistics Management., 20*(3), 360–377.

Grunert, K. G. (2005). Food quality and safety: Consumer perception and demand. *European Review of Agricultural Economics., 32*(3), 369–391.

Hackius, N., & Petersen, M. (2017). Blockchain in logistics and supply chain: Trick or Treat? In *Digitalisation in supply chain management and logistics*. Available at: https://www.econstor.eu/bitstream/10419/209299/1/hicl-2017-23-003.pdf. Accessed on: 24.05.2019.

Hakansson, H., & Persson, G. (2004). Supply chain management: The logic of supply chains and networks. *The International Journal of Logistics Management, 15*(1), 11–26.

Hastig, G. M., & Sodhi, M. S. (2020). Blockchain for supply chain traceability: Business requirements and critical success factors. *Production and Operations Management, 29*(4), 935–953.

Howson, P. (2020). Building trust and equity in marine conservation and fisheries supply chain management with blockchain. *Marine Policy*, 1–6.

Hua, J., Wang, X., Kang, M., Wang, H., & Wang, F (2018). *Blockchain based Provenance for Agricultural Products: A Distributed Platform with Duplicated and Shared Bookkeeping*. In Proceedings of the IEEE Intelligent Vehicles Symposium (IV), Changshu, China.

Kouhizadeh, M., Saberi, S., & Sarkis, J. (2021). Blockchain technology and the sustainable supply chain: Theoretically exploring adoption barriers. *International Journal of Production Economics, 231*, 1–21.

Kshetri, N. (2018). 1 Blockchain's roles in meeting key supply chain management objectives. *International Journal of Information Management, 39*, 80–89.

Leong, C., Viskin, T., & Stewart, R. (2018). Tracing the Supply Chain: How Blockchain Can Enable Traceability in the Food Industry. *Accenture*. Available at: https://www.accenture.com/t20190115T192110Z__w__/us-en/_acnmedia/PDF93/Accenture-Tracing-Supply-Chain-Blockchain-Study-PoV.pdf (Accessed 12 December 2019).

Lin, I. C., Shin, H., Liu, J. C., & Jie, Y. X. (2017). *Food traceability system using blockchain*. In Proceedings of the 79th, IASTEM International Conference, Tokyo, Japan, pp. 59–64.

Lin, J., Shen, Z., Zhang, A., & Chai, Y. (2018). *Blockchain and IoT based food traceability for smart agriculture*. In Proceedings of the 3rd International Conference on Crowd Science and Engineering, Singapore, pp. 28–31.

Lin, W., Huang, X., Fang, H., Wang, V., Hua, Y., Wang, J., Yin, H., Yi, D., & Yau, L. (2020). Blockchain technology in current agricultural systems: From techniques to applications. *IEEE Access, 8*.

Lin, Z., Prajog, D., & Oke, A. (2016). Supply chain technologies: Linking adoption, utilisation, and performance. *Journal of Supply Chain Management, 52*(4), 22–41.

Mai, N., Bogason, S. G., Arason, S., Arnason, S. V., & Matthiasson, T. G. (2010). Benefits of traceability in fish supply chains – Case studies. *British Food Journal, 112*(9), 976–1002.

Pearson, S., May, D., Leontidis, G., Swainson, M., Brewer, S., Bidaut, L., Frey, J. G., Parr, G., Maull, R., & Zisman, A. (2019). Are distributed ledger technologies the panacea for food traceability? *Global Food Security, 20*, 145–149.

Perboli, G., Musso, S., & Rosano, M. (2018). Blockchain in logistics and supply chain: A lean approach for designing real-world use cases. *IEEE Access, 6*, 62018–62028.

Porter, M. E. (1996). What is strategy? *Harvard Business Review*, 61–78.

Post, R., Smit, K., & Zoet, M. (2018). *Identifying factors affecting blockchain technology diffusion*. In Twenty-fourth Americas conference on information systems, New Orleans, pp. 1–10.

Queiroz, M. M., & Wamba, S. F. (2019). Blockchain adoption challenges in supply chain: An empirical investigation of the main drivers in India and USA. *International Journal of Information Management, 46*, 70–82.

Rogerson, M., & Parry, G. C. (2020). Blockchain: Case studies in food supply chain visibility. *Supply Chain Management: An International Journal, 25*(5), 601–614.

Russo-Spena, T., & Mele, C. (2012). "Five Co-s" in innovating: A practice-based view. *Journal of Service Management, 23*(4), 527–553.

Saberi, S., Kouhizadeh, M., Sarkis, J., & Shen, L. (2018). Blockchain technology and its relationships to sustainable supply chain management. *International Journal of Production Research, 57*(7), 2117–2135.

Sander, F., Semeijn, J., & Mahr, D. (2018). The acceptance of blockchain technology in meat traceability and transparency. *British Food Journal, 120*(9), 2066–2079.

Saurabh, S., & Dey, K. (2021). Blockchain technology adoption, architecture, and sustainable Agri-food supply chains. *Journal of Cleaner Production, 284*, 1–13.

Song, L., Wang, X., & Merveille, N. (2020). *Research on blockchain for sustainable E-Agriculture*. In 2020 IEEE Technology & Engineering management conference (TEMSCON).

Stranieri, S., Riccardi, F., Meuwissen, M. P. M., & Soregaroli, C. (2021). Exploring the impact of blockchain o the performance of Agri-food supply chains. *Food Control, 119*, 1–12.

Su, S. I. I., Gammelgaard, B., & Yang, S. L. (2011). Logistics innovation process revisited: Insights from a hospital case study. *International Journal of Physical Distribution & Logistics Management, 41*(6), 577–600.

Tan, B., Yan, J., Chen, S., & Liu, X., (2018). *The Impact of blockchain on Food Supply chain: The Case of Walmart.* In Proceedings of the International Conference on Smart Blockchain 2018, Tokyo, Japan, pp. 166–177.

Tian, F. (2016). *An Agri-food supply chain traceability system for China based on FRID& Blockchain technology.* In 2016 13th International Conference on service systems and service management (ICSSSM), pp. 1–6.

Tian, F. (2017). *A supply chain traceability system for food safety based on HACCP, Blockchain & Internet of Things.* In 2017 International Conference on Service Systems and Service Management. pp. 1–6.

Tieman, M., & Darun, M. R. (2017). Leveraging blockchain technology for halal supply chains. *Islam and Civilisational Renewal., 8*, 547–550.

Tsolakis, N., Niedenzu, D., Simonetto, M., Dora, M., & Kumar, M. (2020). Supply network design to address United Nations Sustainable Development Goals: A case study of blockchain implementation in Thai fish industry. *Journal of Business Research*, 1–25.

Van Hilten, M., Ongena, G., & Ravesteijn, P. (2020). Blockchain for organic food traceability: Case studies on drivers and challenges. *Policy and Practice Review, 3*, 1–13.

Van Hoek, R. (2019). Exploring blockchain implementation in the supply chain. *International Journal of Operations & Production Management, 39*(6/7/8), 829–859.

Verhoeven, P., Sinn, F., & Herden, T. T. (2018). Examples from Blockchain implementations in logistics and supply chain management: Exploring the mindful use of a new technology. *Logistics, 2*(3), 1–19.

Wagner, S. M. (2008). Innovation management in the German transportation industry. *Journal of Business Logistics, 29*(2), 215–232.

Wamba, S. F., Queiroz, M. M., & Trinchera, L. (2020). Dynamics between blockchain adoption determinants and supply chain performance: An empirical investigation. *International Journal of Production Economics, 229*, 1–15.

Yadav, S., & Singh, S. (2020). Blockchain critical success factors for sustainable supply chain. *Resources, Conservation & Recycling, 152*, 1–11.

Yiannas, F. (2018). A new era of food transparency powered by blockchain. *Innovations: Technology, Governance, Globalisation., 12*(1–2), 46–56.

Zhao, G., Liu, S., Lopez, C., Lu, H., Elgueta, S., Chen, H., & Boshkoska, B. M. (2019). Blockchain technology in Agri-food value chain management: A synthesis of applications, challenges and future research directions. *Computer in Industry, 109*, 83–99.

Chapter 9
Implications and Conclusion

This chapter is a summary of the whole research. It provides the answers to the research questions, presents contributions of the research, identifies the research limitations, and provides directions for future research.

Overall, this research has nine chapters: Chap. 1 provides the background of this research with a broad view on food supply chain issues and blockchain potentials. Chapter 2 conducts a systematic literature review, which analysed 58 relevant research papers, and summarises six benefits and four categories of barriers in blockchain implementation. Chapter 3 presents the methodology of this research, including the detailed case design and analysis. Chapters 4–6 are the within-case analysis of the three cases. Chapter 7 conducts a cross-case analysis to compare the three cases and summarise the differences and similarities. Chapter 8 is the discussion chapter that applied the theory lens of the innovation process model and the practice-based view (PBV). By comparing findings from the case companies to the existing literature, seven propositions were proposed and a conceptual framework was proposed as the discussion outcomes. Finally, this chapter provides a conclusion to this research.

9.1 Answers to the Research Questions

This section presents the answers to the two research questions that were proposed in Chap. 1:

Research Question 1): Why do companies apply blockchain in food supply chain management and based on what threats or opportunities in the food supply chain?

Research Question 2): How has blockchain been applied in the food supply chain and how can it influence the food industry?

C. Zhang et al., *Blockchain Applications in Food Supply Chain Management*, Contributions to Management Science,
https://doi.org/10.1007/978-3-031-27054-3_9

9.1.1 Answer to the First Research Question

Food safety is a global issue that can be caused by various reasons. Chapter 2 (Sect. 2.1.2) summarised the characteristics of the food supply chain and found that the food supply chain is extremely complex and fragile compared to other supply chains (Kendall et al., 2019). There are two main reasons for such characteristics. First, unlike other industries, food products are usually perishable with expiry dates, and have strict requirements on storage and delivery conditions (Trienekens & Zuurbier, 2008; Wognum et al., 2011). The quality and safety can be significantly discounted if the products are out of date or stored in a wrong place, particularly for fresh and frozen products. Second, the food supply chain usually involves multiple actors and contains many processes. More importantly, the actors and processes are interconnected with and influential over each other (Stevens, 1989; Folkerts & Koehorst, 1997; Trienekens & Zuurbier, 2008). This means that the supply chain performance and product quality can be impacted by any actor and the processes in between. Moreover, the food supply chain usually needs stakeholders to pay more attention during the processes, as food products can be easily contaminated (Ting et al., 2014). Therefore, to secure food quality and safety, it is necessary to ensure that every step is completed with high standards along the food supply chain with multiple stakeholders.

If incidents happen that require food recall, the traceability system is crucial to reduce negative consequences (Moe, 1998; Opara, 2003, p. 103; Mai et al., 2010; Mattevi & Jones, 2016; Garcia-Torres et al., 2018). With sufficient data records, the traceability system can be used to retrieve information and isolate certain products accurately to reduce unnecessary loss. Thus, it is essential to have an efficient traceability system for food companies to be used as an approach after food safety issues emerge. However, traceability is also suggested to be used as a proactive strategy to prevent food incidents (Moe, 1998; Opara & Mazaud, 2001; Opara, 2003). Food companies, therefore, are motivated to apply various technologies to improve their traceability system, blockchain started to attract increasing attention in recent years (Duan et al., 2020; Pournader et al., 2020; Stranieri et al., 2021).

Since the concept of blockchain was proposed, its business potentials in the supply chain have gradually attracted the interest of researchers and practitioners. Thus, there are a few research explored the potentials of blockchain since 2016 from theoretical perspectives (Tian, 2016, 2017; Kouhizadeh & Sarkis, 2018; Queiroz et al., 2019; Rogerson & Parry, 2020; Stranieri et al., 2021). There are six key benefits and opportunities of blockchain are suggested by the research (Tian, 2016, 2017; Yiannas, 2018; Pearson et al., 2019; Rogerson & Parry, 2020; Stranieri et al., 2021): supply chain and traceability efficiency, Information transparency and supply chain visibility, information authenticity and accountability, supply chain digitisation, supply chain resilience, and sustainability.

From the case findings, the case companies chose blockchain as the solution to address supply chain issues because of blockchain's unique features (Casino et al., 2020; Rogerson & Parry, 2020): decentralisation, immutability, security, and smart contract. For instance, blockchain as a decentralised ledger, can reduce the number of middlemen and promote direct communication between stakeholders. The immutability feature can prevent potential data alteration and ensure data authenticity. Data security ensures that the uploaded data are safe and cannot be tampered with by anyone. Moreover, the tokenisation function allows sensitive data to be changed to a new form of data, which are undecipherable and irreversible. This means that the token cannot be reverted to its original form, and this can largely increase data security and guarantee the uniqueness of the data. The combination of immutability and security can bring to an authentic data record that will speed up the food recall process and help to define liabilities. The smart contract can also be operated on blockchain to facilitate a quick and safe payment process.

From the observations, while deciding to adopt blockchain, the three case companies were facing different threats and opportunities as they had different food supply chain characteristics. For BeefLedger, the fragmented information flow, food quality and authenticity, and inefficient document processing are the key issues for a cross-border beef supply chain. By applying blockchain system, the linear supply chain changes to a circular supply chain due to the blockchain decentralisation feature. Stakeholders, no matter where they are, can share and receive the same information in real time. This can largely streamline the information flow. Blockchain's immutability and security feature can ensure information authenticity. By enabling the smart contract, the documents and certificates can be digitised, and the process for document checking can speed up.

As one of the largest retailers in the world, W company hopes to enhance consumer buying confidence and to improve supplier management efficiency. Similar to BeefLedger, the information immutability and data security features can be applied to ensure data authenticity. By scanning the blockchain QR code, consumers can know more about the stories behind the food products. Sufficient information can help consumers to make educated purchasing decisions. In addition, the decentralisation, immutability, and security feature can also facilitate W company's supplier management. As blockchain is decentralised, suppliers have equal power with W company. Information is secured and permanent, and the data cannot be changed by any party. Therefore, suppliers are encouraged to upload more data to provide consumers with more information. At the same time, W company can receive more data on suppliers' performance for better supplier management. Moreover, liability can also be defined when needed.

In order to create a sustainable coffee supply chain, FairChain aims to achieve an equal value chain, and proves the efforts and results to the public. In this case, blockchain is deployed as a story proving tool because of its immutability feature. By uploading the costs at every stage in the coffee supply chain, FairChain can show how they help local coffee farmers to gain living income. To expand the positive externalities, FairChain also activated the tokenisation function to raise consumer awareness and encourage participation in sustainable behaviours. Consumers, thus, have the chance to invest in the farms by purchasing products, receiving tokens, and activating tokens to invest to farmers.

9.1.2 Answer to the Second Research Question

The innovation adoption process model is applied in Chap. 7 to present the company's blockchain implementation process (Flint et al., 2005; Su et al., 2011). The innovation adoption process model is applied to explain the process of innovative thinking and to present the company activities during decision-making. To answer "how" questions, the innovation process model provides a step-by-step account of how companies make decisions, and what practices have been adopted and valued at the same time.

Companies accomplished the process from idea generation to blockchain implementation within four steps. Based on the observations, sub-constructs are proposed to add more details to the main construct. The first stage is **setting the stage.** At this stage, companies are identifying company issues and demands and inviting business partners. Both this research and existing literature suggest that companies should know their business needs and supply chain issues before selecting technologies, in order to find the right business use case (Verhoeven et al.,2018; Song et al., 2020).

The second stage (**Customer clue-gathering activities**) is to collect data and information that based on the results from the first stage. The data can relate to technology or industry. At this stage, all the case companies adopted several data collection methods, such as conducting internal research, attending industry conferences, and consulting relevant experts. This is the stage to gain as much knowledge and experience as possible, in order to lay a solid ground for the actual implementation. Thus, at this stage, all case companies also either ran pilots (BeefLedger and FairChain) or learned from past piloting experience (W company).

The third stage (**Negotiating, clarifying, and reflecting activities**) is to build the blockchain system and to persuade suppliers to use blockchain. Based on the knowledge and experience accrued from the second stage, companies started to design the blockchain platform, including selecting the base infrastructure, deploying IoT devices, and enabling certain functions. In addition, supplier engagement is found to be one of the key factors to promote blockchain projects both from existing literature and observations (Post et al., 2018; Tan et al., 2018; Hastig & Sodhi, 2020). Thus, case companies adopted different ways to learn suppliers' needs and attract their interests, such as employing professionals to communicate with suppliers, and providing them with blockchain-related training.

The final stage (**Inter-organisational learning**) is to expand the scope of the suppliers and invite more stakeholders, and to optimise the blockchain system and maximise its efficiency. Thus, the rest of the stakeholders including suppliers, customers, and third parties if necessary are expected to participate in the blockchain system. Moreover, case companies also adjusted the blockchain system according to the change in demands and requirements. At this stage, the blockchain system became more mature compared to the previous stage. The stakeholder acceptance also increased gradually. Therefore, companies can enable more functions to achieve better efficiency such as designing the interface to be more user-friendly.

The practice-based view (PBV) is applied to analyse the case companies' blockchain implementation and project outcomes. The PBV is adopted for two reasons: *first*, the PBV can help to find the key elements to success and the common barriers during blockchain implementation. *Second*, the PBV can top up on the innovation process model to provide a complete and deeper view of blockchain implementation.

Three parts were identified: critical success factors, common barriers, and project outcomes. Combined with the existing literature (Post et al., 2018; Hastig & Sodhi, 2020; Yadav & Singh, 2020; Saurabh & Dey, 2021), this research found four key critical success factors and eight sub-factors: capabilities (knowledge, resources), collaboration (leadership, stakeholder acceptance), technology readiness (technology maturity, technology compatibility), and external environment (external support, market awareness).

Capabilities suggest a company's initial condition, such as company size, company financial condition, company knowledge base, and company culture (Hastig & Sodhi, 2020). Thus, knowledge and resources are the two sub-factors under capabilities. Knowledge can refer to a company's understanding of the technology. Resources can include both financial and human resources. The level of understanding of blockchain and resources can influence a company's blockchain adoption.

The importance of **collaboration** in blockchain adoption was mentioned by multiple researchers (Chapman et al., 2002; Hakansson & Persson, 2004; Grawe, 2009; Post et al., 2018; Tan et al., 2018; Dutta et al., 2020; Hastig & Sodhi, 2020). From the observations, leadership and stakeholder acceptance are two main sub-factors. How the case company performs its leadership role can strongly influence stakeholders' acceptance of blockchain. Stakeholder's attitude can directly impact the blockchain project progress, as blockchain implementation needs the participation of multiple supply chain stakeholders.

Technology readiness is a factor drawing on the technological perspective. It contains two sub-factors: the maturity and the compatibility of the technology. Technology maturity level decides whether the market is willing to accept it, as immature technology can be risky to invest in. The compatibility of blockchain refers to whether blockchain can fit in the current company system. Replacing the whole system with blockchain can be costly and risky, particularly for SMEs (Perboli et al., 2018). Thus, if blockchain is sufficiently mature and can be operated conveniently with the existing technologies, this can strongly facilitate blockchain implementation.

External environment can be the influences from the outside. This can include encouragement or pressure from third parties, such as the government. From the observations, all the case companies received different levels of support from governments. The support directly encouraged them to engage with blockchain. For example, BeefLedger and FairChain both received funding from governments on their blockchain projects. Besides, the change of market awareness and reaction can influence a company's decision-making (Rogerson & Parry, 2020). For instance, a growing number of consumers tend to know more information about food products which can encourage food companies to engage with new technology in order to meet consumers' expectations (Rogerson & Parry, 2020; Stranieri et al., 2021).

In addition, common barriers are summarised in four categories among the three cases via the PBV analysis: intra-organisational barriers, inter-organisational barriers, systems-related barriers, and external barriers (Saberi et al., 2018). The **intra-organisational barriers** relate to the focal company itself. For example, SMEs may have concerns over financial constraints while investing in blockchain. The lack of understanding of blockchain also negatively influences blockchain design. From the observation, BeefLedger and FairChain, as start-ups and a SME, had received funding from governments to support their blockchain projects.

The **inter-organisational barriers** are the barriers that are present between stakeholders. Blockchain is an information technology that requires more than one stakeholder on board. Thus, stakeholder engagement also could pose a barrier during blockchain implementation. The case companies provided various incentives to attract stakeholders, such as sufficient education and better shelf place by W company.

The **systems-related barriers** are the barriers of blockchain technology, such as technology immaturity. As blockchain is a young technology with limited use cases, companies are learning by doing; for example, combining on-chain and off-chain data to solve blockchain scalability issues. Deploying IoT devices to facilitate data collection can eliminate raw data errors (Kshetri, 2018; van Hoek, 2019).

The **external barriers** refer to regulations and policies. Currently, the laws and regulations are not sufficient to be applied to blockchain. As the data are permanently recorded on the chain and shared to some users, this leads to concerns about company privacy. Therefore, the lack of sufficient policies is also a barrier that can discourage a company from implementing blockchain.

Moreover, six blockchain benefits are also found from observations that are in line with existing literature (Chap. 2, Sect. 2.4.2.1). The first benefit is that blockchain can improve **supply chain and traceability efficiency** by improving the food recall efficiency and the utilisation efficiency of financial and human resources. For BeefLedger, the food recall speed can be largely improved due to sufficient data records. W company can better manage their suppliers by gaining more supplier performance data. FairChain can encourage consumers to engage in sustainable behaviours by giving tokens, and save money from marketing and advertising.

Information transparency and supply chain visibility can also be improved by blockchain (Rogerson & Parry, 2020). Blockchain's decentralisation feature ensures that all verified users have equal power and share the same information in real time. Thus, BeefLedger can convert the linear supply chain to a circular supply chain. In this way, the supply chain becomes more transparent.

The third benefit is **information authenticity and accountability** due to blockchain's immutability and security feature. As the information is permanent on chain, it is unlikely that it will be changed or that data will be erased during food recall. Thus, the immutable information not only provides consumers with buying confidence but can also help to define responsibilities when needed (Kshetri, 2018; Stranieri et al., 2021).

Supply chain digitisation is another benefit that is brought by blockchain. As data are uploaded on the chain, paper records can be replaced. In addition, IoT devices are also popular to be combined with blockchain platforms to increase data collection efficiency and accuracy. Moreover, smart contract, as one of the most important features of blockchain, can digitise the contract and be executed automatically.

Another benefit is **supply chain resilience**, which refers to how the supply chain performs in risky environments. As verified stakeholders can share real-time information, any mistakes due to fragmented information flow can be largely reduced. This is particularly important for cross-border long supply chains, and for fresh and frozen products. In addition, immutable information records can also help companies to trace products during food recall. It can also detect unethical behaviours of stakeholders such as manipulating data.

Sustainability can also be promoted by blockchain. By increasing supply chain and traceability efficiency, food wastes can be dramatically reduced because companies can better monitor their product flow. Unnecessary wastes due to fragmented information flow or insufficient data record can be eliminated. In the case of FairChain, the tokenisation function can help consumers to participate in the shared value chain and to have more awareness about sustainability.

9.2 Research Contributions

This paper is one of the first to investigate how blockchain influences food supply chain specifically. So far, although research in this area is growing (Ali et al., 2021; Saurabh & Dey, 2021), there is still a shortage of related research, particularly from an empirical perspective. Thus, this research is trying to fill the gap by conducting a systematic literature review and a case study research to provide the results. This research made a few contributions for blockchain research in supply chain management by adopting the innovation process model and practice-based view theories, and for practices (Fig. 9.1 and Table 9.1).

9.2.1 Contributions to Theories

This section presents the research contributions to theory (blockchain research in food supply chain, innovation adoption process model, and practice-based view) and practice (Focal companies, suppliers, and third parties) (Table 9.1).

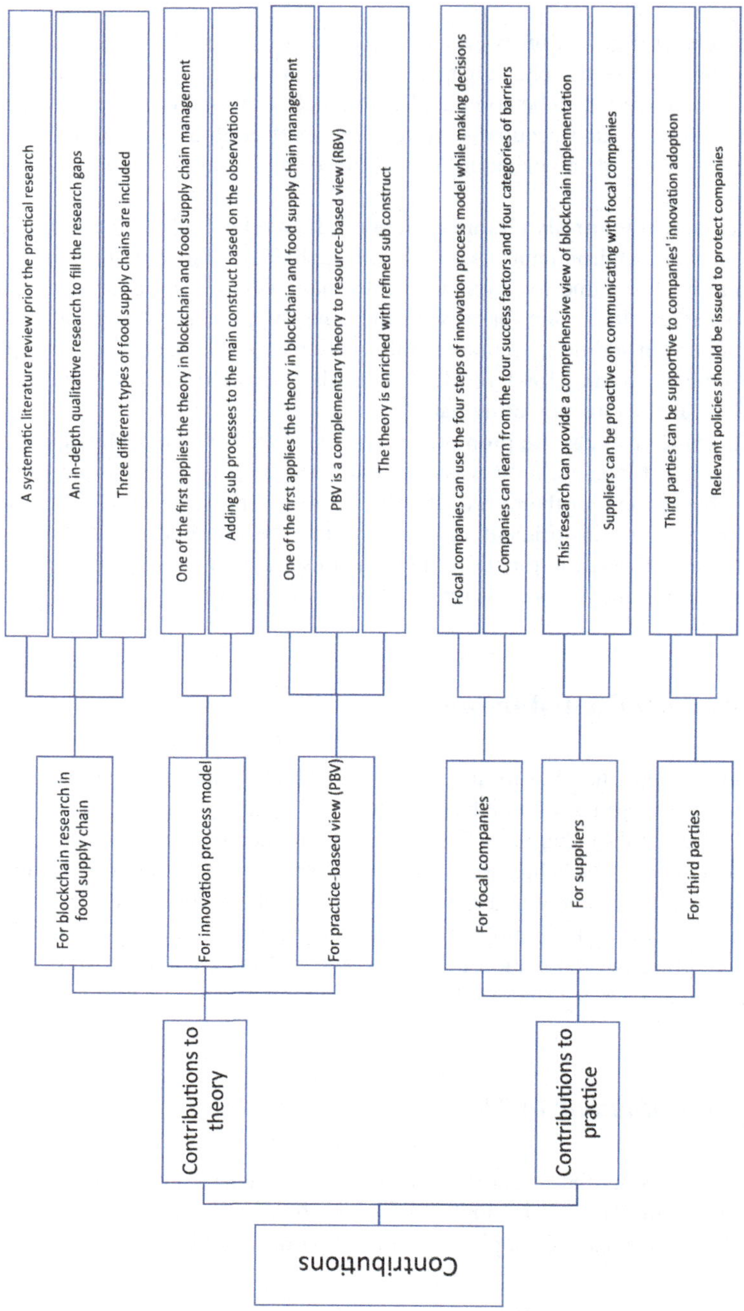

Fig. 9.1 Contributions

Table 9.1 List of contributions

	Contributions
For blockchain research in food supply chain	1. A systematic literature review was conducted that can lay a solid foundation for practical research. 2. This research filled the gap between concept and practice in blockchain research by conducting an in-depth qualitative research. 3. This research investigated three different types of food supply chains in blockchain adoption.
For innovation process model	1. This research is one of the first to apply the innovation process model to blockchain implementation and in food supply chain management. 2. The innovation process model is enriched by adding sub-processes to the main construct based on the observations of the case companies in this research.
For practice-based view (PBV)	1. This research is one of the first that extended the PBV into blockchain implementation in food supply chain management. 2. This research found that the PBV is a complementary theory to the resource-based view (RBV), which agrees with some existing literature. 3. After applying the PBV in blockchain and food supply chain management, the theory is enriched with three refined sub-constructs: critical success factors, common barriers, and project outcomes.
For focal companies	1. This research provides four steps innovation adoption process model that companies can follow while making relevant blockchain implementation decisions. 2. Companies can learn from case companies' practices to pay attention to the four success factors and prevent the four categories of barriers.
For suppliers	1. This research can provide a comprehensive view of blockchain implementation, and the advantages and disadvantages that suppliers need to be aware of before accepting it. 2. Instead of being passive, suppliers can be proactive in communicating with focal companies and participating in blockchain platform development.
For third parties	1. Third parties can also play a critical role in food supply chain management by being supportive to innovation adoption and provide some help. 2. Governments and legislators should issue complete and sufficient policies and regulations to protect companies and provide confidence.

9.2.1.1 Contributions to the Blockchain Research in Food Supply Chain

First, this research provides a fundamental and comprehensive understanding of blockchain and its potential impacts by conducting a systematic literature review. After identifying and analysing the most recent and related papers, the literature review lays a solid ground for the following case study research. Moreover, the systematic literature review also provides a deep understanding and systematic knowledge to either new researchers or practitioners who have strong interests in blockchain and prepares them to explore blockchain in more in-depth research.

Second, this research adopts the case study research method to investigate the company's decision-making process and blockchain implementation process. Compared to existing literature that is mostly conceptual, this research is one of the first attempts to fill the knowledge gap between concepts and practices in blockchain implementation in food supply chain management. By analysing three case studies, this research identifies company decision-making steps, common influencing factors, and barriers.

Third, this research investigates blockchain adoption within three different food supply chains: cross-border beef supply chain, food retail supply chain, and value redistribution of the coffee supply chain. Apart from the different characteristics of the supply chains, the three case-study companies also had different objectives and expressed different concerns while making decisions. However, this research proposes a united conceptual framework that reflects the three supply chains' practices, and seven propositions as the summary of the findings. The propositions include the positive relationships between company preparations (identify issues and demand, have sufficient knowledge, have existing trust with stakeholders, provide sufficient supplier trainers, and be flexible and adaptable to possible system changes) and blockchain implementation, and the positive links between blockchain implementation and business performance. The propositions not only can be a useful guide for practitioners and company decision-makers but also can remind them to use blockchain wisely.

9.2.1.2 Contributions to the Innovation Process Model

First, this research adopts the innovation process model to identify the steps of the company's decision-making process in blockchain adoption. This theory is adopted to answer how companies make decisions including the steps and activities involved during their decision-making processes. Originally, the model was applied under logistics innovation by logistics service providers. Thus, this research further extends the innovation process model under the blockchain and supply chain context. This research agrees with Su et al. (2011) that this model is not only applicable to customers, but also to suppliers. Thus, under the supply chain context, the activities in the third and fourth steps involve multiple stakeholders. The third step (negotiating, clarifying, and reflecting activities) was conducted among suppliers. The final step (inter-organisational learning) was a joint learning stage for all stakeholders, including focal companies, suppliers, and customers. Blockchain is a young technology that most practitioners still have doubts on about how it serves better information flow within a supply chain. Thus, the innovation process model presents how companies value blockchain while making decisions.

Second, four stages (setting the stage activities, customer clue-gathering activities, negotiating, clarifying, and reflecting activities, inter-organisational learning) of the innovation process model were proposed by Flint et al. (2005). This research adopts the four stages as the main constructs to analyse the case companies' decision-making processes but refined the model with added sub-constructs according to the observations. The first stage (Setting the stage activities) involves two sub-processes: Identify the issues and clarify company demand; Identify partners. The second stage (Customer clue-gathering activities) has two sub-processes: Data collection and propose business model; Run pilots. The third stage (Negotiating, clarifying, and reflecting activities) contains two sub-processes: Design and implement new system; Supplier engagement. The fourth stage (Inter-organisational learning) has two sub-processes: Stakeholder collaboration; System adjusting and business adapting. The sub-constructs are summarised and proposed based on the company behaviours. Thus, this research has refined the innovation process model in blockchain adoption.

9.2.1.3 Contributions to the Practice-Based View

First, the PBV was adopted and enriched in this research. The PBV is applied in this research to find the common influencing factors for innovation adoptions. The PBV suggests that companies may encounter similar barriers and adopt similar activities during innovation adoption. Thus, this research is one of the first to apply the PBV in the blockchain context. Thus, the applications of the PBV are extended by this research. This research compares the behaviours of the three case companies and summarises the common and transferable behaviours. By operationalising the theory, this research identifies the three sub-constructs (critical success factors, common barriers, and project outcomes) in food supply chain management.

Second, the theory was originally proposed to find the transferable behaviours between companies. Some suggested it to be opposite to the resource-based view (RBV) (Bromiley & Rau, 2014). However, this research agrees with the opinion that practices usually come with resources (Korkman et al., 2010; Russo-Spena & Mele, 2012). This research finds that the three case companies shared similar resources while implementing blockchain, and faced some common barriers. Thus, the findings suggest that PBV is a complementary theory to the resource-based view (RBV).

Third, the PBV was proposed to find how a company should behave by studying ongoing practices. Thus, this research suggested three perspectives on analysing practices: key success factors (capabilities, collaboration, technology readiness, external environment); common barriers (intra-organisational barriers, inter-organisational barriers, systems-related barriers, external barriers); and project outcomes (supply chain and traceability efficiency, information transparency and supply chain visibility, information authenticity and accountability, supply chain digitisation, supply chain resilience, sustainability). The influencing factors present the key elements that can impact company decision-making and innovation adoption. Common barriers suggest that the barriers emerge from adopting innovations. The project outcomes are the most likely results that companies can expect by practicing similar behaviours. The three perspectives are variable under different scenarios but can provide companies with a general overview of both the positive and negative sides of certain practices.

9.2.2 Contributions to Practice

Apart from theoretical contributions, there are also practical contributions. This section presents how this research can contribute to practitioners, suppliers, and third parties.

9.2.2.1 Contributions for Focal Companies

For companies interesting in adopting blockchain in their food supply chains, this research can provide some deep insights for inspiration, and guidance for practices.

First, this research presents the process of blockchain implementation following the innovation adoption process model. From the initial idea generation to final action, the process is divided into four steps which companies can follow to implement blockchain. The key findings of each step are presented in propositions. For example, first, it is suggested that companies have a specific business need before considering adopt blockchain. Next, a thorough data collection and review of the technology is necessary to lay solid grounds for its implementation. Moreover, collaboration is one of the key factors for blockchain implementation; thus, companies need to ensure their suppliers are educated. Finally, companies should stay flexible for any system updating to fit with supply chain demand.

Second, four key success factors and four categories of barriers are proposed as reminders for companies to pay attention. In other words, apart from what makes blockchain implementation a success, this research also summarises the barriers that companies may face. For instance, collaboration can be both a success factor and a barrier during blockchain implementation. Thus, companies may need to take a leadership role in the project. Moreover, companies can engage suppliers by educating them and covering certain implementation costs. Thus, companies that wish to implement blockchain can follow the guidance that is provided by this research, learn from the case companies' practices, and prevent potential barriers.

9.2.2.2 Contributions for Suppliers

Suppliers can also learn from this research. This research suggests that apart from focal companies, suppliers, as important participants, should also have a deep understanding of the blockchain before accepting it. Sufficient knowledge of the technology value not only can enhance their acceptance and positive attitude to blockchain adoption but can also help to smooth the implementation process when suppliers are eager to learn.

This research also suggests that suppliers should be proactive and keep communicating with the lead companies. Instead of being passive, by expressing their concerns and demands, suppliers can participate in the blockchain platform development and achieve a "win-win" situation. Focal companies can update the blockchain system based on the needs and requirements of the suppliers. Moreover, the research finds that case companies chose to start the blockchain platform with suppliers who either have close relationships with them or suppliers who have stronger interests in and curiosity about blockchain. Thus, it is possible that suppliers with positive attitudes have higher chances to be selected as the pioneers in the project, as well as to build closer relationships with the focal companies.

9.2.2.3 Contributions for Third Parties

For third parties including governments and NGOs, this research can also provide some implications. This research finds that the supportive attitude and encouragement from third parties can greatly help companies to implement blockchain. All the case companies received different levels of support from third parties. For instance, BeefLedger and FairChain applied for funding from governments to launch their blockchain platforms. NGOs and local communities also helped FairChain to solve communication issues with farmers. W company also maintained a close relationship with the government and was invited to the "China Food Safety Publicity Week" for a presentation about blockchain system. Thus, being supportive to the innovations of third parties can greatly enhance a company's confidence in blockchain adoption.

In addition, the barriers that relate to insufficient regulations and policies can also discourage some companies and potential practitioners from trying out blockchain. Thus, to encourage more companies implement blockchain, governments and some legislative institutions may need to issue policies to protect the companies and enhance their confidence about blockchain. Blockchain provides further opportunities to certification standardisation organisations in order them to take part in blockchain and play a critical role.

9.3 Limitations

Despite the notable contributions of this research, there are a few limitations:

First, all three cases have senior managers and key stakeholders who participated in the interviews. However, the supply chain usually has multiple stakeholders. In this research, not all stakeholders were interviewed due to the pandemic from 2020. In addition, due to COVID-19 and international travel restrictions, there were no on-site visits during this research. The pandemic also restricted the researcher from carrying out face-to-face interviews. Thus, all the interviews were carried out by telephone or via online meeting. In future research, face-to-face interviews could help strengthen the depth of interviews and increase the number of interviews.

Second, although, this research has listed many barriers during blockchain implementation, for new adopters or practitioners who have interest in blockchain, there are still risks involved in adopting a technology that is not mature, like blockchain. Because blockchain is a young technology, all the case companies have only adopted it for a limited time. This means that there are still unexpected risks which may happen in the future.

Third, although this research has conducted in-depth interviews with three case companies, the number of case companies is still limited due to multiple reasons. First, blockchain is a young technology, and there are only a limited number of adopters in the food industry. Most of them are still in the piloting phase and still exploring. Therefore, not all participants were willing to share their practices. Second, due to the ongoing pandemic in 2020, many companies have paused or postponed their blockchain projects to reduce investment risks. Therefore, the number of companies that meet the criteria, including meeting certain levels of project maturity and willingness to share experience, has been limited to three in this research.

Fourth, this research identifies the critical success factors and barriers to the case companies. However, different factors and barriers can have different levels of impact on blockchain implementation, particularly for different supply chains. For example, capabilities as key success factors can be the determining implementation factors for SMEs. If SMEs fail to receive support from governments, blockchain projects can be hard to complete. Therefore, future research can investigate and rank different factors and barriers.

9.4 Future Research Directions

By identifying and analysing the most relevant related papers, this work lays a solid foundation for future research directions in this area.

First, future research could focus on specific food products, and present more precise findings. The global food supply chain is vast and complicated. Different food products, such as fresh food, frozen food, agri-food, processed food, and others, may require different supply chain and blockchain solutions. The barriers to and the benefits of blockchain may also be different.

Second, the application can be expanded to other functions under the food supply chain such as supply chain finance, and the recycling chain. Future research can explore how blockchain can be applied in other functions.

Third, future research can look into the performance of blockchain for different aims, such as sustainability, supply chain resilience, and efficiency. When the demands and aims are the same, it is possible to identify the influencing factors for companies to achieve specific goals.

Fourth, this research was conducted and written up during the COVID-19 pandemic. Thus, research on the post-COVID impact on blockchain adoption may be an interesting area. For Customers' requirements regarding food quality and safety may change due to COVID-19, so food companies may also change their product-monitoring systems accordingly.

Fifth, there are gaps about cultural differences on blockchain adoption. Different countries with different cultural backgrounds may have different attitudes to blockchain adoption. For example, the Chinese government shows a supportive attitude and encourages companies to use blockchain to solve certain issues, but is stricter for digital currencies trades.

Sixth, this research explores three fully implemented blockchain cases. Future research may explore cases at different stages of maturity in their blockchain platform. Different stages could encounter different barriers and outcomes. Comparisons can be carried out at a global level or from a within-country perspective.

Seventh, future research can explore the supply chain co-governance mechanisms. It is interesting to find that, in this research, the three blockchains are initiated by the focal firms. According to the different features of blockchain technology, we could expect that the blockchain network will follow a co-governance pattern in the long term which would need further exploration.

Eighth, the supply chain performance can be further studied quantitively in the long run. Such studies can be carried out at the focal firm perspective or the whole supply chain perspective to verify the true value of blockchain technology. Supply chain performance can also be further quantified.

9.5 Final Words

During my 3-year PhD programme, I had a chance to investigate the latest technology—*blockchain*—and explore how it can be applied in food supply chain management. Based on existing literature, my observations of multiple food companies' practices, and my interviews with practitioners, I have completed this research. I hope this research can provide other researchers and practitioners with more thoughts and views about blockchain implementation particularly in its current early stages. I believe that blockchain has a promising future for facilitating better supply chain performance and may disrupt the current supply chain governance. Much more research can be done to expand this promising research domain.

References

Ali, M. H., Chung, L., Kumar, A., Zailani, S., & Tan, K. H. (2021). A sustainable blockchain framework for the halal food supply chain lessons from Malaysia. *Technological Forecasting and Social Change, 170*, 1–14.

Bromiley, P., & Rau, D. (2014). Research prospective: Towards a practice-based view of strategy. *Strategic Management Journal, 35*(8).

Casino, F., Kanakaris, V., Dasaklis, T. K., Moschuris, S., Stachtiaris, S., Pagoni, M., & Rachaniotis, N. P. (2020). Blockchain-based food supply chain traceability: A case study in the dairy sector. *International Journal of Production Research, 59*(19), 1–13.

Chapman, R. L., Soosay, C., & Kandampully, J. (2002). Innovation in logistic services and the new business model: A conceptual framework. *Managing Service Quality, 12*(6), 358–371.

Duan, J., Zhang, C., Gong, Y., Brown, S., & Li, Z. (2020). A content-analysis based literature review in blockchain adoption within food supply chain. *International Journal of Environmental Research and Public Health, 17*, 1–17.

Dutta, P., Choi, T., Somani, S., & Butala, R. (2020). Blockchain technology in supply chain operations: Applications, challenges and research opportunities. *Transportation Research Part E, 142*, 1–33.

Flint, D. J., Larsson, E., Gammelgaard, B., & Mentzer, J. T. (2005). Logistics innovation: A customer value-oriented social process. *Journal of Business Logistics, 26*(1), 113–147.

Folkerts, H., & Koehorst, H. (1997). Challenges in international food supply chains: Vertical co-ordination in the European agribusiness and food industries. *Supply Chain Management: An International Journal, 2*(1), 11–14.

Garcia-Torres, S., Albareda, L., Garcia, M. R., & Seuring, S. (2018). Traceability for sustainability-literature review and conceptual framework. *Supply Chain Management: An International Journal, 24*(1), 85–106.

Grawe, S. J. (2009). Logistics innovation: A literature-based conceptual framework. *The International Journal of Logistics Management, 20*(3), 360–377.

Hakansson, H., & Persson, G. (2004). Supply chain management: The logic of supply chains and networks. *The International Journal of Logistics Management, 15*(1), 11–26.

Hastig, G. M., & Sodhi, M. S. (2020). Blockchain for supply chain traceability: Business requirements and critical success factors. *Production and Operations Management, 29*(4), 935–953.

Kendall, H., Kuznesof, S., Dean, M., Chan, M., Clark, B., Home, R., Stolz, H., Zhong, Q., Liu, C., Brereton, P., & Frewer, L. (2019). Chinese consumer's attitudes, perceptions and behavioural responses towards food fraud. *Food Control, 95*, 339–351.

Korkman, O., Storbacka, K., & Harald, B. (2010). Practices as markets: Value co-creation in e-invoicing. *Australasian Marketing Journal, 18*, 236–247.

Kouhizadeh, M., & Sarkis, J. (2018). Blockchain practices, potentials, and perspectives in greening supply chains. *Sustainability, 10*(10), 1–16.

Kshetri, N. (2018). 1 Blockchain's roles in meeting key supply chain management objectives. *International Journal of Information Management, 39*, 80–89.

Mai, N., Bogason, S. G., Arason, S., Arnason, S. V., & Matthiasson, T. G. (2010). Benefits of traceability in fish supply chains – Case studies. *British Food Journal, 112*(9), 976–1002.

Mattevi, M., & Jones, J. A. (2016). Traceability in the food supply chain: Awareness and attitudes of UK small and medium-sized enterprises. *Food Control, 64*, 120–127.

Moe, T. (1998). Perspectives on traceability in food manufacture. *Trends in Food Science and Technology, 9*(5), 211–214.

Opara, L. U. (2003). Traceability in agriculture and food supply chain: A review of basic concepts, technological implications, and future prospects. *Food, Agriculture and Environment, 1*(1), 101–106.

Opara, L. U., & Mazaud, F. (2001). Food traceability from field to plate. *Outlook an Agriculture, 30*(4), 239–247.

Pearson, S., May, D., Leontidis, G., Swainson, M., Brewer, S., Bidaut, L., Frey, J. G., Parr, G., Maull, R., & Zisman, A. (2019). Are distributed ledger technologies the panacea for food traceability? *Global Food Security, 20*, 145–149.

Perboli, G., Musso, S., & Rosano, M. (2018). Blockchain in logistics and supply chain: A lean approach for designing real-world use cases. *IEEE Access, 6*, 62018–62028.

Post, R., Smit, K., & Zoet, M. (2018). Identifying factors affecting blockchain technology diffusion. *Twenty-Fourth Americas Conference on Information Systems*, New Orleans, pp. 1–10.

Pournader, M., Shi, Y., Seuring, S., & Koh, S. C. L. (2020). Blockchain applications in supply chains, transport and logistics: a systematic review of the literature. *International Journal of Production Research, 58*(7), 2063–2081.

Queiroz, M. M., Telles, R., & Bonilla, S. H. (2019). Blockchain and supply chain management integration: A systematic review of the literature. *Supply Chain Management, 25*(2), 241–254.

Rogerson, M., & Parry, G. C. (2020). Blockchain: Case studies in food supply chain visibility. *Supply Chain Management: An International Journal, 25*(5), 601–614.

Russo-Spena, T., & Mele, C. (2012). "Five Co-s" in innovating: A practice-based view. *Journal of Service Management., 23*(4), 527–553.

Saberi, S., Kouhizadeh, M., Sarkis, J., & Shen, L. (2018). Blockchain technology and its relationships to sustainable supply chain management. *International Journal of Production Research, 57*(7), 2117–2135.

Saurabh, S., & Dey, K. (2021). Blockchain technology adoption, architecture, and sustainable agri-food supply chains. *Journal of Cleaner Production, 284*, 1–13.

Song, L., Wang, X., & Merveille, N. (2020). *Research on blockchain for sustainable E-Agriculture*. 2020 IEEE Technology & Engineering management conference (TEMSCON).

Stevens, G. C. (1989). Integrating the supply chain. *International Journal of Physical Distribution and Materials Management, 19*(8), 3–8.

Stranieri, S., Riccardi, F., Meuwissen, M. P. M., & Soregaroli, C. (2021). Exploring the impact of blockchain o the performance of agri-food supply chains. *Food Control, 119*, 1–12.

Su, S. I. I., Gammelgaard, B., & Yang, S. L. (2011). Logistics innovation process revisited: insights from a hospital case study. *International Journal of Physical Distribution and Logistics Management, 41*(6), 577–600.

Tan, B., Yan, J., Chen, S., & Liu, X. (2018). The impact of blockchain on food supply chain: The case of Walmart. In *Proceedings of the International Conference on Smart Blockchain 2018*, Tokyo, pp. 166–177.

Tian, F. (2016). An Agri-food supply chain traceability system for China based on FRID & Blockchain technology. *2016 13th International Conference on Service Systems and Service Management (ICSSSM)*, pp. 1–6.

Tian, F. (2017). A supply chain traceability system for food safety based on HACCP, Blockchain & Internet of Things. *2017 International Conference on Service Systems and Service Management*, pp. 1–6.

Ting, S. L., Tse, Y. K., Ho, G. T. S., Chung, S. H., & Pang, G. (2014). Mining logistics data to assure the quality in a sustainable food supply chain: A case in the red wine industry. *International Journal of Production Economics, 152*, 200–209.

Trienekens, J., & Zuurbier, P. (2008). Quality and safety standards in the food industry, developments and challenges. *International Journal of Production Economics, 113*(1), 107–122.

Van Hoek, R. (2019). Exploring blockchain implementation in the supply chain. *International Journal of Operations & Production Management, 39*(6/7/8), 829–859.

Verhoeven, P., Sinn, F., & Herden, T. T. (2018). Examples from Blockchain implementations in logistics and supply chain management: Exploring the mindful use of a new technology. *Logistics, 2*(3), 1–19.

Wognum, P. M., Bremmers, H., Trienekens, J. H., & van der Vorst, J. G. A. J. (2011). Systems for sustainability and transparency of food supply chains – Current status and challenges. *Advanced Engineering Information, 25*(1), 65–76.

Yadav, S., & Singh, S. (2020). Blockchain critical success factors for sustainable supply chain. *Resources, Conservation and Recycling, 152*, 1–11.

Yiannas, F. (2018). A new era of food transparency powered by blockchain. *Innovations: Technology, Governance, Globalization., 12*(1–2), 46–56.

Appendices

Appendix 1: Cover Letter for Selected Companies

The Application of Blockchain in Food Supply Chain Management

Research Background

Blockchain has become a hot topic in recent years, and more companies tend to use blockchain to address their company issues. A few blockchain-based applications have been successfully implemented, such as bitcoin. Apart from the financial area, blockchain is found to be beneficial in other areas such as recycling, property management, voting systems, and supply chain management. However, blockchain application in the supply chain is a relatively new, emerging issue.

In this research, I aim to investigate why and how companies choose blockchain over other innovations to improve food supply chain, how to implement blockchain, and what the benefits and challenges are.

Format/Time Scale

Face to face/telephone one-to-one interviews lasting about one hour with key individuals involved in blockchain and food supply chain management.

Project Contact and Interviewer

Miss Chen Zhang, MSc (Southampton, UK), BSc (Plymouth, UK)
PhD researcher
Southampton Business School

C. Zhang et al., *Blockchain Applications in Food Supply Chain Management*, Contributions to Management Science,
https://doi.org/10.1007/978-3-031-27054-3

University of Southampton
Email: cz2g16@soton.ac.uk
Tel: +44 7456662591

Miss Chen Zhang is a PhD researcher specialising in blockchain applications in food supply chain management at the University of Southampton, UK. Chen obtained her Master's in Supply Chain Management and Logistics from the University of Southampton in 2018.

Deliverable

I would like to produce and share a case report and my final research with you.

Sample interview questions:

- What is your understanding of blockchain?
- How does your company link blockchain with supply chain management? Please specify by examples.
- Which department leads blockchain projects internally? What other departments/ functions have been involved and what role do they assume?
- How do your stakeholders (customers and suppliers) accept blockchain? How does your company help them during blockchain implementation?

Appendix 2: Interview Protocol

Interview Protocol for Focal Company

Food Supply Chain-Related Questions

1. What are the issues in the company's supply chain? Can you explain and give examples please?
2. Do the issues mentioned above exist in general in the related industry, or just in your company?
3. What are the possible reasons behind the issues? How can the issues be solved? What strategies need to be adopted to solve the issues?

Blockchain-Related Questions

1. How does your company understand blockchain? What is your company blockchain strategy? How do you position your company's blockchain strategy when compared with others?
2. What motivated you to adopt the blockchain, risk mitigation-orientation or opportunity-orientation? Could you specify with examples?

3. Which department leads the projects internally? What other departments/functions have been involved and what role do they assume? Please specify with real examples.
4. What are the challenges when your company implements blockchain projects? How do you solve them? Please specify by examples.
5. What do you see of blockchain in the future in food supply chain?

Innovation-Related Questions

1. What other innovations has your company adopted or considered adopting? What are the issues that can be fixed by the innovations? Please specify with examples.
2. What kind of innovations are still in use? And what innovations are replaced by blockchain?
3. Compare the innovations and blockchain, what are the key benefits and challenges of blockchain? What makes your company keep using blockchain?

Interview Protocol for Technology Support Team

Blockchain-Related Questions

1. How does your company understand blockchain?
2. How many blockchain projects does your company have in progress so far? How are they doing so far?
3. How have the blockchain projects been designed for the food supply chain? How does it work? Please explain in detail.
4. Where do you see blockchain in the future in the food supply chain?

Food Supply Chain-Related Questions

1. What motivates your clients to adopt blockchain? What are your clients expecting from using blockchain?
2. Do you think blockchain is the right and only "remedy" for your clients' companies?
3. What do you think of your clients adopting blockchain? Do you think they adopt blockchain to fix the issues, or to follow the blockchain "hype"?
4. Do you think you clients have a sufficient level of understanding of blockchain?

Innovation-Related Questions

1. What other innovations do you think can replace blockchain to fix the company issues?

2. Compared to the other innovations, what do you think of the benefits and challenges of blockchain? Are there any benefits or challenges that have been unexpected?

Interview Protocol for Upstream Suppliers

Food Supply Chain-Related Questions

1. What do you think of your current food supply chain? Explain both good aspects and challenges. Give examples and explain in detail.
2. How do you think that innovations may help to improve the food supply chain?

Blockchain-Related Questions

1. How does your company understand blockchain? How was the process of adopting blockchain?
2. How do you think blockchain can change the current food supply chain? Explain in detail.
3. Do you think blockchain can benefit your company? In which way? Explain in detail.
4. Have you received any help with blockchain adoption from focal companies?
5. How do you see blockchain in the future in food supply chains?

Interview Protocol for Downstream Customers

Food Supply Chain-Related Questions

1. Do you have any worries about the current food supply chain? Explain in detail.
2. How do you think that innovations may help to improve food supply chain?
3. How do you see the food supply chain in the future?

Blockchain-Related Questions

1. How do you understand blockchain? How did you hear about blockchain?
2. How do you think that blockchain can change current food supply chain? Explain in detail. Do you think blockchain can well solve your worries mentioned above?
3. How do you see blockchain in the future in food supply chains?